Charles Henry Ralfe

Outlines of Physiological Chemistry

Charles Henry Ralfe

Outlines of Physiological Chemistry

ISBN/EAN: 9783337330712

Printed in Europe, USA, Canada, Australia, Japan

Cover: Foto ©berggeist007 / pixelio.de

More available books at **www.hansebooks.com**

PREFACE.

THIS little work has been compiled in the hope that it may furnish students and practitioners of medicine with a concise and trustworthy laboratory guide to the qualitative and quantitative analysis of the tissues, fluids and excretory products of the human body. It has purposely been made as simple as possible. The best processes have been selected from the English, French, and more particularly from the German text books and journals, and although these processes have been very succinctly described, care has been taken to omit nothing which is essential to success. In the great majority of cases each process has been verified by careful and, in many, by repeated trials.

A short introductory chapter describes the system of classification which is employed throughout the work. It has been thought wise to adhere, with some modification, to the typical formulæ of Gerhardt, instead of adopting the instructive constitutional formulæ which now replace them in the more advanced works on Chemistry. Constitutional formulæ are not as yet in general use and their employment in an elementary work would therefore be apt to occasion great difficulty to the student.

In the appendix the student will find a brief description of the instruments employed and the general methods of procedure in quantitative analysis. This arrangement saves constant reference to small details in the text.

The importance of chemical analysis to medical practice as well as to physiological progress receives fuller recognition every day. Few hospitals are now without physiological and chemical laboratories in which students can obtain the manipulative skill as well as the theoretical knowledge which they require for practical work; and as the number of skilful workers becomes greater, so most certainly will our knowledge develope with it, and become more extensive, more accurate, and above all, more practical.

Lastly, the author must express his deep obligation to his friend Mr. C. Heaton, in whose laboratory most of the work described in these pages was carried out, for many important suggestions which his varied knowledge enabled him to make, and for the kind care he bestowed on the revision of the proof sheets.

LONDON. March, 1873.

CONTENTS.

PAGE

INTRODUCTION.

Definition of Physiological Chemistry—Classification of Organic substances—Hydrocarbon radicals, Alcohols, Acids, Amines, Amides—Synthesis, Analysis . . . .

CHAP.

PART I.
PROXIMATE PRINCIPLES.

I. Saccharine and Amylaceous principles—Glucose, Lactose, Inosite, Starch, Glycogen . . .   1

II. Fatty Principles—*Saponifiable*, Palmitin, Stearin, Olein—*Non-Saponifiable*, Cholesterin, Serolin, Stercorin, Excretin   13

III. Albuminous and Gelatinous principles—*Albumins*, Albumin, Globulin, Fibrin, Derived Albumins, Albuminose, Mucin, Pyin—*Gelatins*, Gelatin, Chondrin, Elasticin, Keratin   24

PART II.
PRODUCTS OF DECOMPOSITION.

IV. Non-Nitrogenous Organic Acids — Monatomic Fatty Acids, Formic, Acetic, Propionic, Butyric, Valeric, Caproic, Capric, Palmitic, Stearic, Oleic—Diatomic Fatty Acids, *Monobasic*, Carbonic, Glycollic, Lactic, Leucic, *Dibasic*, Oxalic, Succinic, Aromatic Acids, Benzoic, Phenol, Resinous Acids, Cholesteric, Cholic, Lithofellic   44

CHAP.   PAGE
V. The Animal Nitrogenous Bases — Monamides,
Glycocin, Leucin, Sarcosin, Cholin, Taurin,
Cystin, Hippuric Acid, Tyrosin, Cerebrin,
Lecithin—Primary Diamides, Urea, Kreatin,
Kreatinin — Secondary Diamides, Uric Acid,
Guanin, Xanthin, Hypoxanthin, Allantoin   .   60

PART III.

INORGANIC CONSTITUENTS.

VI. Inorganic constituents, Estimation of Water and
Total Solids, Separation of inorganic salts from
organic matter—The Alkaline Carbonates, Estimation of Carbonic Acid—Phosphates, Calcium,
Magnesium, Potassium and Sodium Phosphates
—The Sulphates—The Chlorides—Potassium
and Sodium—Lime and Magnesia—Iron—Estimation of Iron—Silicon and Fluorine—Copper, Lead, Arsenic   98

PART IV.   .

TISSUES AND FLUIDS.

VII. Solid Tissues of the Body—Connective Tissue,
Elastic Tissue, Epidermal Tissue, Cartilaginous
Tissue, Osseous Tissue, Dental Tissue, Mus-
cular Tissue, Nervous Tissue   112

VIII. The Digestive Fluids—The Saliva, Oral Digestion,
Gastric Juice, Gastric Digestion, Bile, Pancreatic Juice, Intestinal Juice, Intestinal Digestion .   126

IX. The Circulatory Fluids—The Blood, Absorption
Spectrum of Blood—The Chyle—The Lymph
—Pus   148

CHAP. PAGE

X. The Reproductive Fluids—The Milk—Seminal
Fluid—Fluid of the Mammalian Ovum . . 170

XI. The Excreta—The Breath—The Sweat—The Fæces—The Urine, Urinary Sediments and Calculi 177

APPENDIX I. Weights, Measures and Instruments employed in the Quantitative Analysis of the Tissues and Fluids, and the usual method of procedure 214

APPENDIX II. Standard solutions required in Volumetric analysis 226

APPENDIX III. List of Authors referred to, and from
whose works extracts have been taken . . 229

INTRODUCTION.

THE term *Physiological Chemistry* is generally limited to the study of the chemical phenomena attendant upon the life of Animals.

The objects of study in this branch of science may be summarised as follows.

1. The ultimate and proximate composition of the various parts of the animal body.

2. The nature and composition of the proximate principles of the body, and of the focd from which they are derived.

3. The successive processes by which food is transformed into living tissue, or is otherwise made available for the wants of the body.

4. The processes by which the constituents of the body are removed from it, after fulfilling their proper functions.

5. The development of active force (kinetic energy) in the body, and the production therefrom of heat, electricity, muscular contraction, etc.

COMPOSITION AND CONSTITUTION OF ORGANIC SUBSTANCES. The proximate principles are divided into two distinct groups; viz. 1. The Non-Nitrogenous, and 2. The Nitrogenous. To the first group belong the Saccharine and Oleaginous principles, to the second the Albuminous and Gelatinous.

These principles form the basis of the animal tissues and fluids, and in fulfilling their purposes in the economy are broken up and oxidized. This oxidation is seldom, if ever, accomplished at one stage. Intermediate compounds of simpler constitution are commonly produced, but ultimately the final oxidation is reached, and thus the greater part of the elements of the food are removed from the system in the forms of carbonic acid, water, and urea.

I. PRODUCTS OF THE DECOMPOSITION OF THE SACCHA-
RINE AND OLEAGINOUS PRINCIPLFS.

Lactic Acid
Oleic   ,.
Stearic   ,,
Palmitic ,,
Butyric   ,,
Acetic   ,,
Formic   ,,
Oxalic   ,.
Carbonic ,,
Water

II. PRODUCTS OF THE DECOMPOSITION OF THE ALBU-
MINOUS PRINCIPLES.

Non-nitrogenous bodies.	*Nitrogenous bodies.*
Lactic Acid	Xanthin
Oleic ,,	Hypoxanthin
Stearic ,,	Cystin

Palmitic ,,	Uric Acid
Butyric ,,	Hippuric Acid
Acetic ,,	Leucin
Formic ,,	Cholin
Oxalic ,,	Tyrosin
Carbonic ,,	Kreatin
Water	Urea

In the above tables we see that the Saccharine and Oleaginous principles break up into a single series of non-nitrogenous fatty acids, the lowest term of which is Carbonic acid. The Albuminous principles by their decomposition furnish a double series; one of which is identical with the products of decomposition furnished by the saccharine and fatty bodies, whilst the other consists of certain nitrogenous bodies or amides, the lowest term of which is urea, the ammoniated form of carbonic acid.

There is no essential difference between organic and inorganic chemistry. ' Organic chemistry is simply the chemistry of carbon compounds, and accordingly we find that the proximate principles that we meet with in the animal body consist of carbon, united in various proportions with hydrogen, oxygen, nitrogen, and some other less abundant elements, such as sulphur, phosphorus, and iron.

Many of the more simple of these compounds may be referred for convenience to the three following types two of which will frequently be used in the following pages.

Hydrogen.[•]   Water.   Ammonia.

$\left.\begin{array}{l} H \\ H \end{array}\right\}$ $\left.\begin{array}{l} H \\ H \end{array}\right\} O.$ $\left.\begin{array}{l} H \\ H \\ H \end{array}\right\} N.$

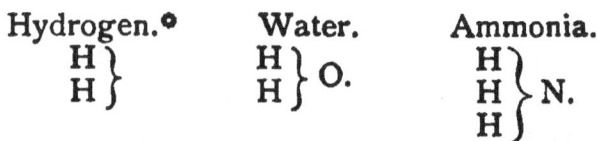

HYDROCARBONS.

(Compounds of Carbon and Hydrogen.)

The classification of these bodies is based upon the atomicity of carbon, which being a tetrad element requires 4 atoms of hydrogen, or some other monad, for its full saturation, i.e., to satisfy all its combining powers ; the fully satisfied hydrocarbon molecule will therefore be represented by the formula CH_4. Each additional atom of carbon requires, however, only two additional atoms cf hydrogen to maintain the saturation, because a portion of the combining power of each carbon atom is employed in linking the carbon atoms together. The following diagram illustrates this important theory, which it must be remembered is applicable to all kinds of carbon compounds.

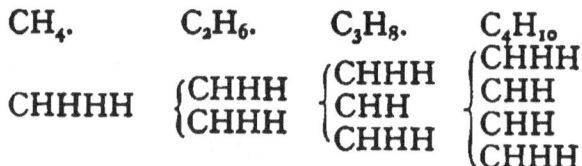

CH_4. C_2H_6. C_3H_8. C_4H_{10}

CHHHH $\left\{\begin{array}{l} CHHH \\ CHHH \end{array}\right.$ $\left\{\begin{array}{l} CHHH \\ CHH \\ CHHH \end{array}\right.$ $\left\{\begin{array}{l} CHHH \\ CHH \\ CHH \\ CHHH \end{array}\right.$

It follows from this that all carbon compounds arrange themselves in series, the members of which differ from one another by CH_2 or by some multiple of CH_2. Thus we have formic acid, CH_2O_2; acetic acid, $C_2H_4O_2$; propionic acid,

---

[•] The hydrogen type is seldom employed in the present day.

$C_3H_6O_2$; and so on. Series of this kind are termed *homologous series*, and the members are said to be *homologues* of one another.

Hydrocarbon radicals. From the above considerations, we deduce as the general formula for a saturated hydrocarbon the expression $C_n H_{2n+2}$, in which n may denote any number of atoms. If a hydrocarbon contains a less number of hydrogen atoms than the above formula requires, it is, or may be, a *radical*; that is it may exist in compounds, and play therein the part of an elementary atom.°

The *atomicity* of such a radical must obviously depend on the number of hydrogen, or other monad atoms required to complete it. Thus the radical CH_3 is a monad; CH_2 a diad; and CH a triad. Their function in compounds is well illustrated by their chlorides which are strictly comparable to metallic chlorides.

Chloride of

Sodium NaCl. Zinc   Zn''Cl$_2$. Bismuth Bi''Cl$_3$

Methyl CH_3Cl. Methylene CH_2Cl_2. Formyl $CHCl_3$

Those of the hydrocarbon radicals which contain an *even* number of hydrogen atoms are capable of existing in the separate state, and most, but not all of them, have actually been prepared. Ethylene C_2H_4, and Acetylene C_2H_2, are examples. Those, on the other hand, which contain an *uneven*

° Some hydrocarbons which appear from their formulæ to be unsaturated are really saturated, or at any rate have an atomicity less than that indicated by the above theory.

It is not necessary in this place to enter into the theoretical explanation of this apparent anomaly.

b

number, such as methyl CH_3; ethyl C_2H_5; and glyceryl C_3H_5, cannot exist in the free state, but only in compounds.

The names and formulæ of a few of the more important hydrocarbon radicals are given in the ollowing table.

TABLE OF PRINCIPAL HYDROCARBON RADICALS.

MONADS. Methyl Series.	DYADS. Olefine Series.
Methyl CH_3	✿
Ethyl C_2H_5	Ethylene C_2H_4
Propyl C_3H_7 ·	Propylene C_3H_6
Butyl C_4H_9	Butylene C_4H_8
Amyl C_5H_{11}	Amylene C_5H_{10}
Hexyl C_6H_{13}	Hexylene C_6H_{12}

TRIADS. Glycerin Series.
✿
Ethine or Acetylene C_2H_2
Propine or Allylene C_3H_4
Quartine or Crotonylene C_4H_6
Quintine or Valerylene C_5H_8
Sextine or Diallyl C_6H_{10}

Hydrocarbons of the Aromatic series. Benzene C_6H_6 and *Toluene* C_7H_8 are the most important members of this series. By the rule before given they should be octads, but they possess the properties of saturated hydrocarbons, and are therefore not to be reckoned among the radicals. From them are derived the important monad radicals *phenyl* C_6H_5 and *toluyl* C_7H_7.

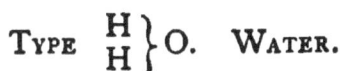

$$\text{Type} \quad \left.{H \atop H}\right\} O. \quad \text{Water.}$$

The constitutions of the alcohols, ethers, aldehydes, and organic acids, can be represented simply by regarding them as derivatives of one or more molecules of water $\left.{H \atop H}\right\} O,\quad \left.{H_2 \atop H_2}\right\} O_2,\quad \left.{H_3 \atop H_3}\right\} O_3$ etc. ;

I. *Alcohols.* In an alcohol a monad, diad, triad, etc., hydrocarbon radical replaces one or more atoms of H in one or more molecules of ¡water. It is in fact a hydrate of a hydrocarbon radical, as the following examples show.

(a) Ordinary ethyl alcohol C_2H_6O is formed by the monatomic radical ethyl C_2H_5 replacing 1 atom of H in the single molecule of water $\left.{H \atop H}\right\} O$ thus,

$$\left.{C_2H_5 \atop H}\right\} O.$$

(b) Glycerin or glyceryl alcohol $C_3H_8O_3$ is formed by the triatomic radical glyceryl C_3H_5''' replacing 3 atoms of H in the treble molecule of

water $\left.\begin{array}{l}H_3 \\ H_3\end{array}\right\} O_3$ thus, $\left.\begin{array}{l}C_3H_5''' \\ H_3\end{array}\right\} O_3$.

(c) Mannite $C_6H_{14}O_6$, a saturated hexatomic alcohol, is formed by the hexatomic radical C_6H_8 replacing 6 atoms of hydrogen in the molecule $\left.\begin{array}{l}H_6 \\ H_6\end{array}\right\} O_6$

thus, $\left.\begin{array}{l}C_6H_8 \\ H_6\end{array}\right\} O_6$.

II. *Aldehydes*. If an alcohol be submitted to oxidation, it loses 2 atoms of H,* and is converted into a neutral body, called an *aldehydes*, which having a great affinity for oxygen rapidly absorbs it from the air and is converted into an acid. Aldehydes are therefore compounds intermediate between the alcohols and the acids.

(a) Ethyl alcohol C_2H_6O deprived of 2 atoms of H forms *ethyl aldehyde* C_2H_4O, and ethyl aldehyde by oxidation yields *acetic acid* $C_2H_4O_2$.

(b) Mannite $C_6H_{14}O_6$ deprived of two atoms of H forms *mannitose* $C_6H_{12}O_6$, a sugar isomeric with glucose, and mannitose by oxidation yields *mannitic* acid $C_6H_{12}O_7$.

III. *Organic Acids*. As stated above the organic acids may be regarded as alcohols, in which a portion of the hydrogen of the radicals is replaced by oxygen. They are therefore formulated as derived from a single, double, or treble molecule of water by the replacement of H, H_2 or H_3 by a monad, diad, or triad oxygenated hydrocarbon radical.

---

* Hence its name, *alcohol de hydrogenatus*.

Examples. (a) Acetic acid $C_2H_4O_2$ is formed by the oxidized radical *acetyl* C_2H_3O which has replaced 1 atom of H in water, thus, $\left.\begin{array}{c} C_2H_3O \\ H \end{array}\right\} O.$

(b) Glycollic acid, $C_2H_4O_3$, is a double molecule of water in which half the hydrogen is replaced by *glycollyl* $\left.\begin{array}{c} C_2H_2O \\ H_2 \end{array}\right\} O_2.$ In fact it is ethylene alcohol $\left.\begin{array}{c} C_2H_4 \\ H_2 \end{array}\right\} O_2,$ in which two atoms of the hydrogen of ethylene are replaced by one atom of oxygen.

(c) Oxalic acid, $C_2H_2O_4$, is a double molecule of water in which half the hydrogen is replaced by *oxalyl* $\left.\begin{array}{c} C_2O_2 \\ H_2 \end{array}\right\} O_2.$ Here the whole of the hydrogen of the ethylene in ethylene alcohol is replaced by oxygen. It will be seen that both these acids are related to ethylene alcohol as acetic acid is to ethyl alcohol.

———

A tabular arrangement of the principal alcohols and acids and the hydrocarbon radicals from which they are derived is given in the next page.

A Table of the Principal Alcohols and Acids.

FATTY GROUP I.

| | | ACID. | |
RADICAL.	ALCOHOL.	Monobasic.	Dibasic.
Methyl Series or Monatomic.			
CH_3	Methyl CH_4O	Formic CH_2O_2	
C_2H_5	Ethyl C_2H_6O	Acetic $C_2H_4O_2$	
C_3H_7	Propyl C_3H_8O	Propionic $C_3H_6O_2$	
C_4H_9	Butyl $C_4H_{10}O$	Butyric $C_4H_8O_2$	
C_5H_{11}	Amyl $C_5H_{12}O$	Valerianic $C_5H_{10}O_2$	
C_6H_{15}	Hexyl $C_6H_{14}O$	Caproic $C_6H_{12}O_2$	
C_8H_{17}	Octyl $C_8H_{18}O$	Capric $C_8H_{16}O_2$	
$C_{16}H_{33}$	Cetyl $C_{16}H_{34}O$	Palmitic $C_{16}H_{32}O_2$	
$C_{18}H_{37}$	Stearyl $C_{18}H_{38}O$	Stearic $C_{18}H_{36}O_2$	
Ethine Series or Diatomic.			
C_2H_4	Ethylene $C_2H_6O_2$	Carbonic CH_2O_3	Oxalic $C_2H_2O_4$
C_3H_6	Propylene $C_3H_8O_2$	Glycollic $C_2H_4O_3$	Malonic $C_3H_4O_4$
C_4H_8	Butylene $C_4H_8O_2$	Lactic $C_3H_6O_3$	Succinic $C_4H_6O_4$
C_5H_{10}	Amylene $C_5H_{12}O_2$	Butylactic $C_4H_8O_3$	Pyrotartaric $C_5H_8O_4$
C_6H_{12}	Hexylene $C_6H_{14}O_2$	Valerolactic $C_5H_{10}O_3$	Adipic $C_6H_{10}O_4$
		Leucic $C_6H_{12}O_3$	

Series		Alcohol	Acids
Glycerine Series or Triatonic.	C_3H_8 * * *	Glycerine $C_3H_8O_3$ * * *	Acrylic $C_3H_4O_2$ Damaluric $C_7H_{12}O_2$ Damolic $C_{13}H_{24}O_2$ Oleic $C_{18}H_{34}O_2$ †
Hexatomic Series.	C_6H_8	Mannite $C_6H_{14}O_6$	Mannitic $C_6H_{12}O_7$

Aromatic Group II.

	Alcohol	Acids
C_6H_6	Phenyl C_6H_6O	Benzoic $C_7H_6O_2$
C_7H_8	Benzyl C_7H_8O	Cinnamic $C_9H_8O_2$
C_9H_{12}	Cinnyl‡ $C_9H_{10}O$	

* These bodies have not yet been isolated.

† These acids for convenience will be described, olęic with the monatomic fatty acids, damaluric and damolic with the aromatic acids.

‡ Cholesterin, $C_{26}H_{44}O$, is homologous with this compound, and possesses in fact some of the properties of an alcohol.

$$\text{Types } \left.\begin{array}{c} H \\ H \\ H \end{array}\right\} N. \text{ Ammonia.}$$

Amines. When a hydrocarbon radical replaces the typical hydrogen of the molecule $\left.\begin{array}{c} H \\ H \\ H \end{array}\right\} N.$ the resulting compound is called a primary, secondary or tertiary *amine*, according as one, two, or three atoms of hydrogen are replaced. Thus the following amines are obtained by the substitution of the hydrogen of ammonia by methyl.

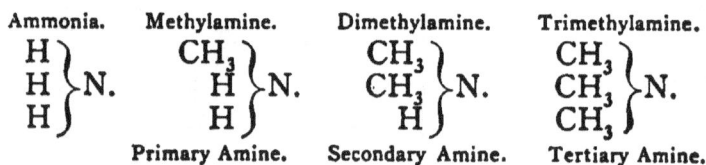

Ammonia.	Methylamine.	Dimethylamine.	Trimethylamine.
$\left.\begin{array}{c} H \\ H \\ H \end{array}\right\} N.$	$\left.\begin{array}{c} CH_3 \\ H \\ H \end{array}\right\} N.$	$\left.\begin{array}{c} CH_3 \\ CH_3 \\ H \end{array}\right\} N.$	$\left.\begin{array}{c} CH_3 \\ CH_3 \\ CH_3 \end{array}\right\} N.$
	Primary Amine.	Secondary Amine.	Tertiary Amine.

Amides. When an acid radical replaces any part of the typical hydrogen of ammonia the resulting compound is called an *Amide.* As some of the amides play a very important part in the animal economy it is necessary to study their constitution a little more closely.

For this purpose it will be convenient to write the formulæ of a few important acids in a form which is a slight variation of that previously used.

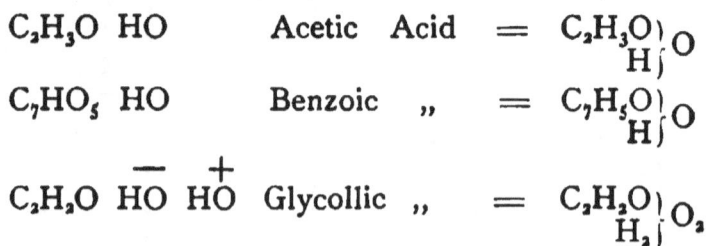

C_2H_3O HO Acetic Acid $= \left.\begin{array}{c} C_2H_3O \\ H \end{array}\right\} O$

C_7HO_5 HO Benzoic „ $= \left.\begin{array}{c} C_7H_5O \\ H \end{array}\right\} O$

C_2H_2O $\overset{-}{HO}$ $\overset{+}{HO}$ Glycollic „ $= \left.\begin{array}{c} C_2H_2O \\ H_2 \end{array}\right\} O_2$

$$C_2O_2 \quad \overset{+}{HO} \ \overset{+}{HO} \quad \text{Oxalic} \quad \text{,,} \quad = \quad \left.\begin{matrix}C_2O_2\\H_2\end{matrix}\right\}O_2$$

$$C_3H_4O \quad \overset{-}{HO} \ \overset{+}{HO} \quad \text{Lactic} \quad \text{,,} \quad = \quad \left.\begin{matrix}C_3H_4O\\H_2\end{matrix}\right\}O_2$$

$$C_3O_3 \quad \overset{+}{HO} \ \overset{+}{HO} \quad \text{Mesoxalic ,,} \quad = \quad \left.\begin{matrix}C_3O_3\\H_2\end{matrix}\right\}O_2$$

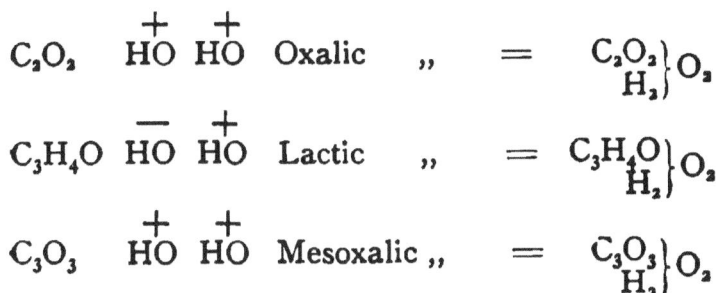

In the above table we have acids of three different kinds, all capable of yielding amides.

1. In acetic and benzoic acids we have examples of acids which are simply monobasic. The amides of these acids are called monamides and are very simple.

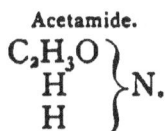

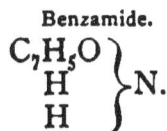

Acetamide.

$$\left.\begin{matrix}C_2H_3O\\H\\H\end{matrix}\right\}N.$$

Benzamide.

$$\left.\begin{matrix}C_7H_5O\\H\\H\end{matrix}\right\}N.$$

They differ from the corresponding amines, just as the acids do from the alcohols.

2. Oxalic and mesoxalic acids are examples of dibasic acids. They are, in fact, dihydrates of the radicals C_2O_2 and C_3O_3. Now these radicals, being diads, are capable of replacing two atoms of hydrogen in the double molecule of ammonia. In this way neutral amides of the kind called *diamides* are formed.

Urea is by far the most important of the diamides.

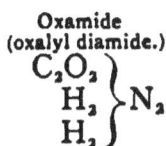

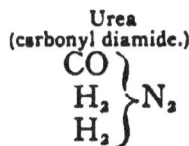

Oxamide
(oxalyl diamide.)

$$\left.\begin{matrix}C_2O_2\\H_2\\H_2\end{matrix}\right\}N_2$$

Urea
(carbonyl diamide.)

$$\left.\begin{matrix}CO\\H_2\\H_2\end{matrix}\right\}N_2$$

But from all dibasic acids a *monad* as well as a diad radical may be derived by merely deducting HO. Thus from sulphuric acid, SO_2 HO HO, we get not only the diad radical SO_2, but also the monad radical SO_2 HO. The following formulæ exhibit this.

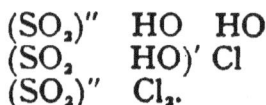

$$(SO_2)'' \quad HO \quad HO$$
$$(SO_2 \quad HO)' \; Cl$$
$$(SO_2)'' \quad Cl_2.$$

When one of these *monatomic* radicals of a *dibasic* acid replaces the hydrogen of ammonia, a mona-mide is formed. But as there is still one atom of replaceable hydrogen attached to the radical, this atom can at any time be replaced by a metal or a hydrocarbon radical and thus the acid character is not entirely lost. From a dibasic, the acid becomes, in fact, a monobasic one. Acids of this kind are called *Amic Acids.* Thus we have.

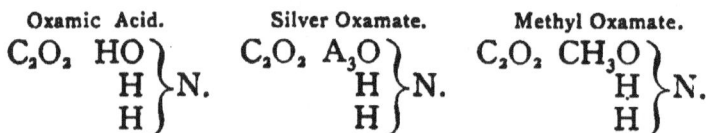

Oxamic Acid.	Silver Oxamate.	Methyl Oxamate.
C_2O_2 HO $\left.\begin{array}{c} \\ H \\ H \end{array}\right\} N.$	C_2O_2 A_2O $\left.\begin{array}{c} \\ H \\ H \end{array}\right\} N.$	C_2O_2 CH_3O $\left.\begin{array}{c} \\ H \\ H \end{array}\right\} N.$

3. Glycollic and lactic acids are examples of acids which are *diatomic* as to their structure and *monobasic* as to their properties. Only one of the two typical hydrogen atoms that each contains can be replaced by metals. The difference has been indicated in the formulæ given above for the acids by marking the replaceable hydrogen by a *plus* and the non-replaceable, by a *minus* sign. The effect of this peculiarity is that *two monad* radicals, one neutral and one acid, can be derived

from each acid. These radicals replace one atom of hydrogen in the single molecule of ammonia just as the monad radicals of dibasic acids do, and monamides are formed; but these monamides are amic acids if the radical so introduced contain the replaceable atom of hydrogen, or neutral amides if it contain only the non-replaceable atom. Thus from glycollic acid we have;

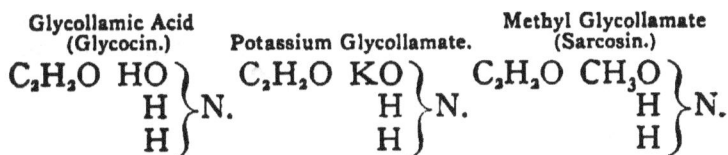

Glycollamic Acid (Glycocin.) Potassium Glycollamate. Methyl Glycollamate (Sarcosin.)

$$\left.\begin{array}{l} C_2H_3O\ HO \\ H \\ H \end{array}\right\}N. \qquad \left.\begin{array}{l} C_2H_3O\ KO \\ H \\ H \end{array}\right\}N. \qquad \left.\begin{array}{l} C_2H_3O\ CH_3O \\ H \\ H \end{array}\right\}N.$$

and also;

Glycollamide.

$$\left.\begin{array}{l} C_2H_3O\ HO \\ H \\ H \end{array}\right\}N.$$ Which does not yield salts.

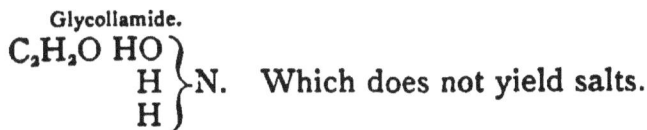

Analogous to this last we have *leucin*, the neutral amide of leucic acid, which is one of the homologues of glycollic acid. The following formulæ describes it.

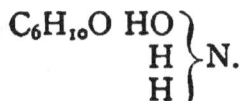

$$\left.\begin{array}{l} C_6H_{10}O\ HO \\ H \\ H \end{array}\right\}N.$$

In other cases a second of the atoms of the hydrogen in ammonia is replaced by an acid radical. Thus *hippuric acid* only differs from glycocin by having one hydrogen atom replaced by the radical of benzoic acid.

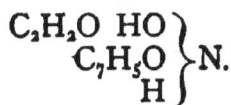

$$\left.\begin{array}{l} C_2H_3O\ HO \\ C_7H_5O \\ H \end{array}\right\}N.$$

Derivations of Urea.

A somewhat numerous and complex class of
bodies is known, the members of which contain
radical, or *residues* of urea together with radicals
derived from various acids. Many of the derivatives of uric acid belong to this class, as also do
the important compounds *kreatin* and *kreatinin.*
In many cases great difference of opinion exists
as to the exact structure of these compounds as
they may be described by several formulæ. Kreatin is generally considered as containing residues
of urea and sarcosin, (methyl-glycocin). Its formula on the ammonia type may therefore be
written as follows.

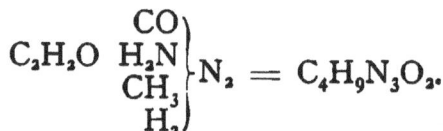

$$C_2H_2O \left. \begin{matrix} CO \\ H_2N \\ CH_3 \\ H_2 \end{matrix} \right\} N_2 = C_4H_9N_3O_2.$$

It is capable of taking up the elements of water
and splitting into urea and sarcosin. The following comparison of the formulæ of these two compounds will serve to illustrate this. The atoms
which have to be removed to produce kreatin are
printed in italics.

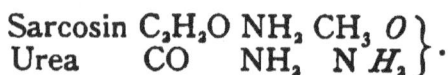

Sarcosin C_2H_2O NH_2 CH_3 O }
Urea CO NH_2 N H_2 } .

Compounds which contain one urea radical are
called *monureides.* Kreatin is a monureide and so
are *paraban* and *alloxan* which are obtained by the
oxidation of uric acid.

Paraban.	Radical of Oxalic Acid.	Radical of Urea.

$$\left.\begin{array}{l} CO \\ C_2O_2 \\ H_2 \end{array}\right\} N_2 \quad = \quad C_2O_2 \qquad CO\ H_2N_2$$

Alloxan.	Radical of Mesoxalic Acid.	Radical of Urea.

$$\left.\begin{array}{l} CO \\ C_3O_3 \\ H_2 \end{array}\right\} N_2 \quad = \quad C_3O_3 \qquad CO\ H_2N_2$$

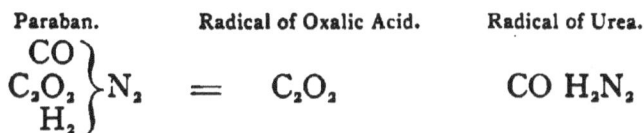

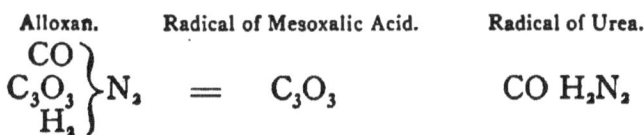

We therefore understand why, when alloxan is boiled with baryta water, it takes up a molecule of water and yields mesoxalic acid and urea.

Compounds which contain two urea radicals are called *diureides*. *Allantoin, xanthin hypoxanthin*, and uric acid itself belong to this class. There is some doubt as to exact constitution of uric acid. It is often represented as consisting of one radical of tartronic acid and two of urea.

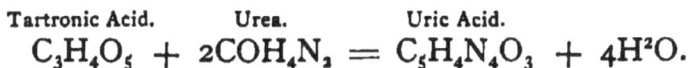

Tartronic Acid.	Urea.	Uric Acid.

$$C_3H_4O_5 + 2COH_4N_2 = C_5H_4N_4O_3 + 4H^2O.$$

This however is merely hypothetical.

SYNTHESIS AND ANALYSIS.

For many years it was supposed that organic substances could only be formed by the agency of a living organism. In 1828 however Wöhler obtained urea by evaporating ammonium cyanate, and since that time chemists have obtained by artificial means a large number of compounds formerly obtainable only from animal or vegetable organisms. These syntheses are effected either by bringing together molecules of simpler constitution to form a more complex body, as in the case of hippuric acid;

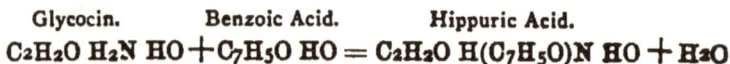

Glycocin.   Benzoic Acid.   Hippuric Acid.

$C_2H_2O \ H_2N \ HO + C_7H_5O \ HO = C_2H_2O \ H(C_7H_5O)N \ HO + H_2O$

or by building up an organic compound from purely inorganic sources; as Berthelot obtained formic acid, by heating carbon monoxide with potassium hydrate at 100° C.

Carbon monoxide.   Potassium hydrate.   Potassium formate.

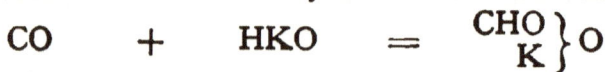

$$CO \quad + \quad HKO \quad = \quad \left. \begin{matrix} CHO \\ K \end{matrix} \right\} O$$

or as Kolbe formed acetic acid from carbon disulphide.

In nature this formation of organic compounds from inorganic materials *(synthesis)* is effected by the agency of the Vegetable Kingdom. The plant under the influence of the rays of the sun liberates a quantity of oxygen from inorganic constituents such as carbonic acid, water, and ammonium carbonate, which exist in the soil and air, converting them into those saccharine, oleaginous, and albuminous principles which form its tissues and juices, and which ultimately furnish the Animal world with food. For example, carbonic acid by deoxidation under certain conditions may yield mannite; thus

$$6CO_2 \quad + \quad 7H_2O \quad - \quad 13O = C_6H_{14}O_6.$$

That plants exert this deoxidizing power is readily shown by a very simple experiment. If a bunch of fresh green leaves be plunged into a broad necked bottle containing fresh spring water,

or water containing carbonic acid in solution, the
bottle turned mouth downwards into a basin of
water so as to exclude the air, and the whole
placed in strong sunlight for an hour or more,
the leaves will become covered with minute bubbles of oxygen, which is derived from the decomposition of the carbonic acid of the water;
the oxygen being set free, while the carbon is
absorbed by that plant to form its tissues.   In this
process of deoxidation however a considerable
quantity of force derived from the sun's rays is
rendered latent, or to speak more accurately becomes potential, one portion being taken up by
the liberated oxygen, the other accumulated in the
tissues and juices of the plant; and this force will
remain latent till the oxygen and carbon are
again united.

The reunion of carbon with oxygen is effected,
either by the direct burning of carbon in oxygen,
as takes place when fuel is burnt in our grates or
stoves; or when the carbon elements of food and
tissues are submitted to the action of the respired
oxygen, and the potential energy assumes the
active or *kinetic* condition of Heat and Motion,
whilst the carbon compounds are reduced again to
their original inorganic state, of carbonic acid,
water, and ammonium carbonate.   That animals
exhale carbonic acid is demonstrated by the turbidity produced by passing a current of expired
air through lime water; the lime being converted
into calcium carbonate or chalk; and we know
that oxygen is absorbed, from the fact of its dimin-

ution in the atmosphere of close, crowded, and ill-ventilated apartments.

Thus we see that the vegetable organism is chiefly employed in building up, *synthetically*, inorganic into organic matter, whilst the animal, *analytically* reduces organic compounds back again to their original inorganic coustituents. These processes of deoxidation and oxidation, do not at once raise or reduce the various substances to or from their elementary condition. On the contrary the process is slow and gradual, several intermediate products being formed. De Luca for example has shown that in the ripening of the olive, carbonic acid is first replaced by certain acids, as oxalic, tartaric etc, and these by mannite; which in its turn is deoxidixed and converted into olein. On the other hand in the decomposition of albumin a number of intermediary bases and fatty acids are formed, such as uric acid, xanthin, kreatin, lactic acid, and oxalic acid before its final oxidation to urea and carbonic acid.

PHYSIOLOGICAL CHEMISTRY.

PROXIMATE PRINCIPLES.

CHAPTER I.

THE SACCHARINE AND AMYLACEOUS PRINCIPLES.

SACCHARINE.		AMYLACEOUS.	
GLUCOSE.	$C_6H_{12}O_6$	STARCH.	$C_6H_{10}O_5$
LACTOSE.	$C_{12}H_{24}O_{12}$	GLYCOGEN.	$C_6H_{10}O_5$
INOSITE.	$C_6H_{12}O_6$		

THESE substances are also termed *Carbo-hydrates*, from the fact of their containing Hydrogen and Oxygen in proportion to form water, united with Carbon; owing to this composition they are readily converted the one into the other by the removal or addition of the elements of water; thus,

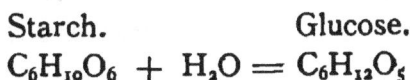

$$\text{Starch.} \qquad\qquad \text{Glucose.}$$
$$C_6H_{10}O_6 + H_2O = C_6H_{12}O_5$$

The researches of Berthôlet have shewn that these substances are alcohols, or are related to the alcohols, of some of the higher poly-atomic radicals; thus Mannite, $C_6H_{14}O_6$, which is a well-defined example of the saccharine series,

is regarded as an alcohol formed by the hexatomic radical C_6H_8 replacing 6 atoms of H, in
the molecule $\left.\begin{array}{c} H_6 \\ H_6 \end{array}\right\} O_6$ to form $\left.\begin{array}{c} C_6H_8 \\ H_6 \end{array}\right\} O_6$; and
glucose is isomeric with mannitose $C_6H_{12}O_{61}$, an
aldehyde formed by the removal of 2 atoms of
H from mannite $C_6H_{14}O_6$.

The carbo-hydrates closely resemble one another in their chemical characters; they are neutral in their reaction, and have little disposition
to enter into combination; they all have a strong
action on polarized light. The saccharine principles (with the exception of inosite) reduce alkaline
copper solutions, throwing down the cuprous
oxide; boiled with liquor potassæ, they give a
brown colour to the solution; and their solutions
undergo vinous fermentation when yeast is added.
The amylaceous principles are converted into the
saccharine by the action of dilute acids; treated
with free iodine they form coloured compounds,
that of starch yielding a characteristic deep blue
colour.

GLUCOSE. $C_6H_{12}O_6$ (syn. Grape Sugar.)

Grape sugar is present as a normal constituent
in healthy blood; it is also found in fœtal urine,
and in the fluid contents of the amnion and allan-
tois; it is most probably present in extremely
minute quantities in normal adult urine; in the
disease known as diabetes a large quantity is always found in the urine.

Sugar fulfils the following purposes in the economy; 1. By its decomposition into carbonic acid and water it fnrnishes a certain quantity of heat which aids in maintaining the temperature of the body: 2. Before its ultimate conversion into these products is completed, it furnishes by its oxidation a number of fatty acids of which the most important is lactic acid. These acids set free in the circulation are essential for the performance of the functions of the body; for without them no acid phosphates* could be formed, nor would there be sufficient acid generated for the digestion of albuminous substances; 3. Sugar also is converted into fat in the body, the fatty acids formed by its oxidation probably uniting with glycerin to form the neutral fats or glycerides.

Sugar is introduced into the economy with the food; either directly with the saccharine principles, or indirectly by the transformation of the starchy matters by the action of the salivary, pancreatic, and intestinal secretions. Also there can be little doubt that sugar is formed in the organism itself by the conversion of *glycogen*, a starchy principle found abundantly in muscular and liver tissue, into glucose. (See glycogen.)

Preparation. · Glucose can readily be obtained,

---

* The acid phosphates are formed by the decomposition of the neutral phosphates; thus,

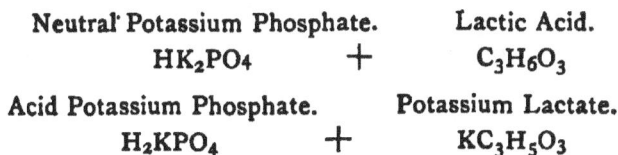

Neutral Potassium Phosphate.		Lactic Acid.
HK_2PO_4	$+$	$C_3H_6O_3$
Acid Potassium Phosphate.		Potassium Lactate.
H_2KPO_4	$+$	$KC_3H_5O_3$

when present in an organic mixture, by evaporating the liquid to dryness on a water bath and pouring hot water on the residue; the aqueous solution will contain sugar, chlorides, if present must be removed by precipitation with silver nitrate, and the filtered solution concentrated to a thick syrup, when the glucose will crystallise out.

Chemical and physical properties. The crystals usually form irregular warty conglomerations, which under the microscope exhibit rhombic tablets; they are soluble in their own weight of water, less soluble in alcohol, and completely insoluble in ether. The aqueous solution is not so sweet as a solution of cane sugar of similar strength would be; it turns the plane of polarization to the right.

Yeast added to a solution of grape sugar, and the mixture kept at a temperature of 30°C, speedily induces vinous fermentation.

Albuminous ferments added to grape sugar solutions, induce lactic and subsequently butyric acid fermentation.

Solutions of grape sugar boiled with liquor potassæ acquire a brownish tint.

If a solution of grape sugar, to which a few drops of liquor potassæ and dilute cupric sulphate solution have been added, be heated, the copper salt is reduced, and a red precipitate of cuprous oxide is thrown down. (As other substances, such as uric acid, kreatin, also have a reducing action on the copper salt, it is well in the case of urine to filter that liquid through animal charcoal before applying the test).

Grape sugar boiled a few minutes with an ammoniacal solution of silver nitrate reduces the metal, which is deposited as a mirror like coating.

Concentrated solutions of sodium chloride and grape sugar deposit flat elongated crystals ; such crystals are obtained on evaporating diabetic urine.

(For quantitative estimation of sugar, see Urine.)

LACTOSE. $C_6H_{12}O_6$ (syn. Milk Sugar).

Lactose is a characteristic constituent of milk, and is elaborated by the mammary glands from the glucose of the maternal blood. It is found in no other secretion. The large quantity always present in milk is an evidence of the important part which sugar takes in the nutrition of the young mammal. The following table shows the per centage quantity in the milk of different animals.

Average quantity of Lactose in 100 parts of milk.

Colostrum	7.	Goat	4.
Human	6.	Ewe	4.
Cow	5.	Sow	3.
Ass	5.	Bitch	1·8.

Preparation. Curdle some milk by the addition of dilute sulphuric acid, and separate the curds by filtration; evaporate the filtrate to the crystallizing point: the crystals thus obtained are very impure, and must be redissolved, purified by filtration through animal charcoal, and recrystallized.

Chemical properties. The crystals form large four-sided prisms with pointed ends, they are extremely hard and have a faint sugary taste. They are insoluble in alcohol and ether, but soluble in 5 pts. of cold water and in dilute acetic acid. Its solutions turn the plane of polarized light to the right. Like glucose it reduces copper salts from their alkaline solutions and silver salts from their ammoniacal solutions; it differs from glucose in not readily undergoing vinous fermentations when yeast is added.

Boiled for some hours with dilute acids it forms *galactose,* a sugar isomeric with glucose, and like it, ferments readily on the addition of yeast. Galactose treated with nitric acid yields mucic acid.

Quantitative estimation of lactose in Milk. Curdle, say 100 C.C. of fresh milk and remove the curds by filtration; dilute 10 C.C. of the filtrate with distilled water to the volume of 200 C.C. and with this fill a Mohr's burette; for the remainder of the process proceed as directed in the quantitative estimation of diabetic urine. (See Urine.)

INOSITE. $C_6H_{12}O_6$ (syn. Muscle sugar.)

Inosite was originally discovered by Scherer in the muscular substance of the heart. It is a constant constituent of voluntary muscular fibre, and is also found in the tissues of the kidney, liver, spleen, brain, and testicle. In certain diseases, as Diabetes, Bright's disease, Phthisis, Syphilitic cachexia and Typhus, it is met with in the urine, an evidence of the incomplete oxidation effected in the system in these diseases.

Preparation. Treat a quantity of muscular tissue as directed in the preparation of kreatin, (See kreatin); and when the kreatin has crystallized out, concentrate the mother liquor, precipitate with dilute sulphuric acid, and remove the precipitate by filtration; the filtrate is then to be heated and ' shaken with ether, to remove the volatile fatty acids; the etherial solution having been poured off, alcohol must be added till a turbidity is produced. The liquid must now be allowed to stand some time, and the precipitate that forms removed by filtration; to the filtrate more alcohol must be added when another precipitate, consisting of crystals of potassium sulphate mixed with crystals of inosite, is thrown down; the latter are to be picked out dissolved in warm water and re-crystallized. (Scherer's process).

Chemical and physical properties. The crystals contain 2 molecules of water; and assume two forms: 1. The most common, are arranged in cauliflower-like groups consisting of oblique prisms, 2. The other variety, are right rhombic prisms. Heated to 100° they lose their water of crystallization and form a white, efflorescent mass of anhydrous inosite. The crystals are soluble in 6 parts of cold water, but are insoluble in alcohol

and ether.   The aqueous solution is sweet but not
syrupy, and keeps unaltered for some time.   In
contact with decaying albuminous matter, inosite
undergoes lactic and subsequently butyric acid
fermentation.

The following is a characteristic test for inosite;
a small quantity of the fluid containing inosite is
evaporated to dryness on a piece of platinum foil
with a drop or two of nitric acid, the residue moistened with ammonia and calcium chloride, which on
evaporation yields a beautiful rose tint.

Although inosite is isomeric with glucose, it pre-
sents none of the characteristic reactions of the
group; thus it does not reduce the copper and silver salts from their alkaline solutions, nor does it
undergo vinous fermentation with yeast, nor has
it any action on polarized light.

STARCH. $C_6H_{10}O_5$.

Starch entering the animal economy in the food
is speedily converted into glucose by the action of
the salivary, pancreatic, and intestinal juices.   In
certain parts of the body as in the prostate, in the
ependyma of the ventricles, the fornix, the choroid
plexus, the retina, and spinal cord, starch granules,
the "corpora amylacea" of Kolliker and Purkinje,
are some times found in advanced age ; these
bodies vary in size from extremely minute granules of $\frac{1}{1800}$ of an inch in diameter, which refract
light strongly, to bodies about one or two lines in
diameter formed by the conglomeration of smaller

· granules. They give a deep blue colour with tincture of iodine, but sometimes the addition of sulphuric acid is required to develope the reaction.

These bodies are not to be confounded with the so called amyloid substance or lardacein.

Chemical and physical properties. Starch occurs in small, rounded granules of irregular form, marked with concentric laminæ, and having a pore "the hilum" at one spot on its surface. The granules boiled in water swell up, burst, and form a stiff mass or paste; they are insoluble in alcohol and ether. Treated with dilute sulphuric acid, starch is converted into 2 molecules of dextrin and 1 molecule of glucose. Heated to 160° starch is converted into dextrin. Starch is dissolved by nitric acid, and on the addition of water to the solution a white substance "xyloidin" is thrown down. Starch with free iodine forms a deep blue compound, which loses its colour when heated to 100° but regains it on cooling.

GLYCOGEN. $C_6H_{10}O_5$ (Syn. Animal Starch).

Glycogen is found in the tissue of muscle and liver, also in the placenta and embryonic tissues of all animals. It is most probably, chiefly derived from the carbo-hydrates introduced into the body with the food, since the largest amount of glyco-gen is obtained from the liver after a diet formed exclusively of these substances; at the same time there can be little doubt that the decomposition of albuminoid substances, which are broken up in the hepatic cells into certain nitrogenous products on the one hand, and into non-nitrogenous products on the other, furnish among the latter class a proportion of the glycogen met with in the liver; and this supposition is rendered more probable since the liver of animals fed exclusively on meat still yield a considerable quantity of glycogen.

Opinions are divided among physiologists as to the final changes glycogen undergoes in the body. Some maintain that it passes from the liver unaltered into the circulation, and is there reduced into carbonic acid and water. Others consider that glycogen is transformed into sugar immediately after its formation in the liver, and in this state is carried off by the hepatic vein into the system; the recent investigations of Dr. Dalton have strengthened the latter view. That gentleman, having removed the liver from a living animal and rapidly comminuted it, found sugar to be always present (the average proportion being 2·59 parts in 1000) within six seconds after the commencement of the experiment. Other experimenters have shown that the blood of the hepatic vein, obtained from living animals, is richer in sugar than the blood coming to the liver. These facts show that sugar is formed by the liver during life and is not a *post-mortem* change as is urged by some writers; for although there is a great increase in the quantity of the sugar in the liver after death, still this increase may be accounted for by supposing that the conversion of glycogenic matter into sugar goes on as long as any glycogen remains in the liver tissue, whilst the sugar which is formed from it is no longer carried away by an active circulation. Moreover, glycogen is easily converted into sugar by any of the animal ferments, such as those contained in the saliva or in the blood. From these considerations Dr. Dalton maintains that "the formation of sugar in the liver " is a function composed of two distinct and suc- " cessive processes ; viz., first, the formation in " the hepatic tissue of a glycogenic matter; and " secondly, the conversion of this glycogenic mat- " ter into sugar by a process of catalysis and " transformation."

Glycogen is also formed in muscular tissue, probably as one of the products of the decomposition of the albuminous constituents ; and by its own decomposition evidently furnishes a supply of force, which helps to maintain the energy of the muscular contractions, since muscular action is associated with a marked diminution in the amount of glycogen in muscular tissue. (Wiess).

Preparation. The liver taken from an animal just killed is rapidly comminuted, and thrown into boiling water for a few minutes, the coagulated mass is then removed, drained, and triturated in a mortar, and boiled for a quarter of an hour; the solid mass is then removed by filtration. The filtrate is then concentrated, and set aside to cool ; and when cold, a drop of hydrochloric acid and a drop of potassio mercuric iodide solution is alternately added, stirred thoroughly for 5 minutes and filtered.

To the filtrate alcohol is added in minute quantities till a precipitate appears, when no more alcohol is to be added lest other substances be precipitated. The precipitate is then removed to a weighed filter and washed with dilute alcohol (60 per cent.), and afterwards with glacial acetic acid; it is then to be dried and weighed. (Brücke's method).

Chemical and physical properties. Glycogen is a yellowish white, amorphous, substance; soluble in water, insoluble in alcohol. With iodine it gives a violet or maroon red coloration. Boiled with dilute hydrochloric acid, it is converted into dextrine and glucose; mixed with saliva at temperatures of 36°, it is converted into glucose. It does not reduce copper salts from their alkaline solutions.

PROXIMATE PRINCIPLES.
CHAPTER II.

___

THE FATTY PRINCIPLES.

SAPONIFIABLE.		NON-SAPONIFIABLE.	
PALMITIN.	$C_{51}H_{98}O_6$	CHOLESTERIN.	$C_{26}H_{44}O$
STEARIN.	$C_{57}H_{110}O_6$	SEROLIN.	$\left.\begin{array}{c} \\ \\ \end{array}\right\}$?
OLEIN.	$C_{57}H_{104}O_6$	STERCORIN.	
		EXCRETIN.	$C_{78}H_{156}SO_2$

The fatty principles are divided into the saponifiable and the non-saponifiable.

THE SAPONIFIABLE OR NEUTRAL FATS are formed by the union of a fatty acid radicle with glycerin; thus, stearin consists of three parts of the acid radical, stearyl $C_{18}H_{35}O$, which has replaced three atoms of the typical hydrogen from $\left.\begin{array}{c} C_3H_5 \\ H_3 \end{array}\right\}$ O_3 glycerin, to form the new substance $\left.\begin{array}{c} C_3H_5 \\ 3(C_{18}H_{35}O) \end{array}\right\}$ O_3 or stearin; and in a similar manner the radicals of palmitic and oleic acid unite with glycerin to form palmitin $\left.\begin{array}{c} C_3H_5 \\ 3(C_{16}H_{31}O) \end{array}\right\}$ O_3, and olein $\left.\begin{array}{c} C_3H_5 \\ 3(C_{18}H_{33}O) \end{array}\right\}$ O_3. Human fat is formed of a mixture of stearin, palmitin, and olein; the two former constitute about three-fourths of the fat of the body and form the solid portion; whilst the olein represents the remaining fourth, and is the liquid or oily constituent.

These substances belong to the fixed oils; they are neutral bodies, of soft greasy consistence, lighter than water. Heated with alkalies, they are *saponified*, *i. e.* the fatty acid unites with the alkali to form a soap, whilst the glycerine remains in solution: this process however takes place only in the presence of water. They are soluble in ether, benzol, fluid oils, carbon bisulphide, chloroform, and hot alcohol, but insoluble in water. They are highly inflammable; heated to 280° in the air, they are decomposed with the formation of acrolein, (if olein is present sebacic acid is also formed). In the living body these fatty substances are liquid at the ordinary temperatures of the blood, but on removal from the body stearin and palmitin crystallize out from the fluid olein.

Fat is found in all the tissues and fluids of the body, usually forming distinct masses or globules, which do not combine with the other elements of the body, but remain free, either suspended in fluids, or lodged between fibres, or deposited in cells.

Fat is essential to the growth and nutrition of the tissues; a larger proportion of fat being met with wherever cell growth is going on rapidly, than in tissues which are fully developed. Fat by its combustion in the economy furnishes a quantity of force to maintain the energy, and heat to supply the temperature of the body; indeed fat may be regarded as the storehouse of carbon; and one apparent advantage of its freedom from combination with other elements is that it is always ready for immediate service, whenever the requirements of the system demand it.

The chief source of the fat of the tissues is of course from the oleaginous constituents of the food; but fat is also formed by the decomposition of the saccharine and albuminous principles, which yield fatty acids, and which, probably combining with glycerin, are converted into fat before their ultimate reduction to carbonic acid and water.

Fat is found in only extremely minute quantities in the healthy human excretions, since in the body it is always decomposed into carbonic acid and water, and in this form passes out of the economy. In certain diseases however fat appears in the excretions. In cases of occlusion, for instance, of the pancreatic and biliary ducts the fats introduced with the food into the intestinal canal are not emulsified and consequently are not absorbed, but pass unaltered out of the system with the fæces. Again, in chyluria, a disease apparently the result of defective oxidation, fat globules are met with in the urine.

Whenever the process of oxidation is impeded or imperfectly performed, we find that fat accumulates in the organs and tissues; for example, we find it in the fatty degeneration of the liver and voluntary muscular fibre met with in all cases of phthisis, or pulmonary diseases which have run a chronic course; and in obesity, the penalty of sedentary or self-indulgent habits. Whenever the supply of blood is cut off from a part, or its flow diminished, the oxidation of that part is of course arrested, and, as a consequence, fatty degeneration occurs; for instance, a thrombus blocks up a cerebral artery, and acute softening of the cerebral substance supplied by that artery is the result. And even if the supply be only diminished instead of entirely arrested, the result is the same only not so rapid. If an organ or member is long disused, or its functional activity impaired, it undergoes fatty degeneration, since the physiological stimulus being no longer supplied, the same quantity of blood does not circulate through the part as when it was in full activity and vigour; the fatty degeneration of the muscular fibres of the uterus after delivery, and the tendency to accumulate fat after the active work of life is over, illustrate this point.

Certain poisons, as the salts of the bile acids, phosphorus, etc., when introduced into the system produce rapid fatty degeneration of the organs and tissues, by causing the destruction of the blood corpuscles and the consequent diminution of the oxidizing power of the blood.

Quantitative estimation of fatty matters. Evaporate a definite quantity of the fluid or tissue (finely divided) in a platinum capsule to dryness, pulverize the dry residue, and agitate with ether, till the residue is thoroughly exhausted, then pour off the etherial solution, and evaporate it to dryness in a *weighed* platinum capsule: the increase in weight of the capsule gives the amount of fatty matter present in the quantity of fluid or tissue examined.

STEARIN. $C_{57}H_{110}O_6$.

Can be obtained by melting mutton suet in a water bath, and adding an equal quantity of ether; the mixture is stirred for some time, and the etherial solution decanted, which when cold will deposit stearin, this must be purified by frequent recrystallization. Stearin occurs in white crystalline nodules which are tasteless and neutral, melting point 69·7° (Heintz) but this point is subject to considerable modification. They are soluble in boiling alcohol and ether, from which they are deposited on cooling.

PALMITIN. $C_{51}H_{98}O_6$.

This substance is obtained from human fat by separating it from the stearin with which it is combined; melting point 62·8°. According to the researches of Heintz, the substance known as *margarin* consists of 10 per cent. of stearin and 90 per cent. of palmitin. Margarin is obtained by heating fat in a water bath, and stirring with an equal quantity of alcohol for some time; the alcoholic solution is then filtered off, and on cooling it will deposit delicate needle-shaped crystals, which arrange themselves in whorled groups or feathers. The melting point of margarin is 47·8°, which is considerably lower than the melting point of its constituent components.

c

OLEIN. $C_{57}H_{104}O_6$.

Is a colourless oil, without taste or smell, re-
maining· liquid at 0° C., exposed to the air it
absorbs oxygen, and becomes rancid; heated in
air to 280° C. it is decomposed with the formation
of sebacic acid. Nitrous acid converts olein into
elaidin. Olein can be obtained tolerably pure by
heating fat in a flask, and filtering it when cold ;
the filtered solution is concentrated by evaporation,
and by the addition of water the olein is separated ;
the product is then exposed to cold at 0° C., and the
solid portion submitted to pressure, the liquid that
separates is almost pure olein.

OLEOPHOSPHORIC ACID is a yellowish gummy substance, composed of oleic acid, glycerin, and
phosphoric acid. It is obtained by treating the
etherial extract of brain substance with ether ; and
to this etherial solution adding some dilute sulphuric acid to remove the alkaline bases with
which the oleophosphoric acid is combined ; the
excess of acid is removed by repeated washing
with water and the etherial solution evaporated.
The dry residue being dissolved in boiling alcohol,
the alcoholic solution on cooling deposits oleophosphoric acid, which must be purified by repeated
washings with ether and alcohol. Oleophosphoric
acid is insoluble in cold absolute alcohol. Agitated with alkaline solutions it is decomposed,
yielding oleates, phosphates, and glycerin. Boiled
for some hours in water, it separates, on cooling,
into two layers, the upper one containing olein,

the lower phosphoric acids. Oleophosphoric acid in combination with soda occurs in all parts of the body, but chiefly in the large nervous centres and voluntary muscles of vertebrate animals.

THE NON-SAPONIFIABLE fatty matters are distinguished from the preceding by not being decomposed or saponified when treated with alkaline solutions. Consequently they can be separated from the other fats by adding a solution of caustic potash to the etherial solution, which causes the saponifiable fats to dissolve out, leaving the non-saponifiable in solution.

CHOLESTERIN. $C_{26}H_{44}O$.

This substance is generally regarded as an excretory product formed in the substance of the brain and nervous tissue, whence it is absorbed by the blood and carried to the liver; here it is separated from the blood and discharged with the bile into the intestines where it undergoes decomposition. Dr. Flint has observed that cases of severe structural disease of the liver are often accompanied by symptoms of blood poisoning, and he attributes this to the accumulation of cholesterin in the blood, or cholesteræmia.

Cholesterin can be obtained from nearly all the tissues and fluids of the body; it also occurs in several morbid products, as gall-stones, the fluid of hydatid, ovarian cysts, etc.

c 2

PROPORTION OF CHOLESTERIN IN 100 PARTS OF

Nervous tissue	3·0	Blood serum .	0·15
Bile	0·25	Blood globules	0·04

Preparation. The fluid or tissue, from which cholesterin is to be extracted, must be thoroughly exhausted with ether; the etherial solution, is then filtered, agitated with a solution of caustic potash, to remove the saponifiable fats, and evaporated; the dried residue treated with boiling alcohol, which on cooling deposits the cholesterin.

Chemical and physical characters. Cholesterin is a white, crystalline substance, somewhat resembling spermaceti; deposited from an alcoholic solution, it forms characteristic, glistening, rhombic plates, which are lighter than water, and melt at 137°. Cholesterin is soluble in ether and boiling alcohol, slightly soluble in oil of turpentine, insoluble in water and cold alcohol.

Touched with a drop of strong nitric acid, and gently evaporated, cholesterin gives a yellow colour, which turns red on the addition of a drop of ammonia.

A mixture of 2 parts of strong hydrochloric acid and 1 part of ferric chloride slightly diluted, evaporated with cholesterin, gives a beautiful violet-coloured residue.

Heated with caustic potash cholesterin gives off hydrogen gas.

Heated with nitric acid *cholesteric acid* $C_8H_{10}O_5$ is formed.

Heated with sulphuric acid and water in equal parts, certain hydrocarbons are formed called *cholesterilines* (Zwenger).

. Heated with dilute phosphoric acid, hydrocarbons termed *cholesterones* are formed (Zwenger).

No rational formula has as yet been assigned to explain the constitution of cholesterin,[*] but it is probable, as suggested by Gerhardt, that it is a monobasic alcohol; this view is strengthened by the fact that cholesterin heated with various hydrated acids in sealed tubes forms certain cholestearin ethers or cholesterides; thus

Cholesterin.   Stearic Acid.   Stearo-cholesterin.

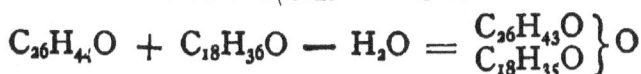

$$C_{26}H_{44}O + C_{18}H_{36}O - H_2O = \left.\begin{matrix} C_{26}H_{43}O \\ C_{18}H_{35}O \end{matrix}\right\}O$$

just in the same manner that stearic acid unites with glycerine to form stearin glycerin.

SEROLIN.

Is a substance obtained, originally by Boudet, by exhausting dried blood serum with ether or boiling alcohol. According to Gobley, this substance is not a simple body but a mixture of different fats. (See also stercorin).

STERCORIN. Under this name Dr. Austin Flint has described a substance which, if not identical with serolin, resembles it closely in its physical and chemical characters. It appears to be a characteristic constituent of fæcal matter, and Dr. Flint

[*] Cholesterin is homologous with cinnyl alcohol $C_9H_{10}O$, of the aromatic series.

assumes that under ordinary circumstances about
0·6 grm. is excreted daily.

Preparation. Dr. Flint directs the fæces to be
evaporated to dryness, pulverized, and exhausted
with ether. The etherial solution is then passed
through animal charcoal, fresh ether being added,
until the original quantity of ether extract has
passed through. The filtered etherial solution is
then evaporated, and the residue treated with boiling alcohol. The alcoholic solution is evaporated,
and the residue treated with a warm solution of
caustic potash to dissolve out all the saponifiable
fats. The mixture is then diluted with water,
thrown on a filter, and washed till the droppings
are clear and neutral. The filter is dried, and the
residue washed out with ether. The etherial
solution is then evaporated, and the residue treated
with boiling alcohol, the residue of this solution
yielding stercorin.

Chemical and physical characters. Stercorin, when
first obtained, appears as a clear, amber-coloured,
oily substance, in which thin, needle-shaped crystals, frequently arranged in bundles, and having
their borders split longitudinally, appear in the
course of a few days. Stercorin is neutral, soluble in ether and hot alcohol, insoluble in water
and solutions of potash; it is distinguished from
cholesterin by having a lower melting point, viz.
38°C. Treated with strong sulphuric acid it gives
a red colour. Dr. Flint considers that stercorin
is formed by a modification of cholesterin in its
passage along the intestinal canal; since a com-

parison of the total quantity of cholesterin contained in bile with the quantity of stercorin actually
discharged shows a correspondence.

EXCRETIN. $C_{78}H_{156}SO_2$.

This principle was obtained by Dr. Marcet,
together with excretolic acid, from fæcal matter. The fæces are first dried and exhausted
with boiling alcohol, and the alcoholic solution
concentrated, filtered, and allowed to stand;
after some time a granular, olive coloured, fatty
acid, excretolic acid is deposited. This substance melts at 25°, is insoluble in water and
in solutions of potash, and in cold alcohol; its
composition has not yet been determined. The
excretolic acid must be removed by filtration, and
the filtrate treated with milk of lime, which throws
down a brown precipitate; this is dried and exhausted with ether, which yields crystals of excretin. The crystals form delicate, silky, four-sided
prisms, insoluble in water, and solutions of potash,
very soluble in ether; they melt at 95° and have
an alkaline reaction.

CHAPTER III.

THE ALBUMINOUS AND GELATINOUS PRINCIPLES.

Albumins.	*Pyin.*
Albumin.	Gelatins.
Globulin.	*Gelatin.*
Fibrin.	*Chondrin.*
Derived Albumins.	*Elasticin.*
Albuminose.	*Keratin.*
Mucin.	

The albuminoids are the most important of the proximate principles, constituting as they do the nutritive element, and the basis of all the tissues and fluids of the body. Thus they are found in white of egg, in the blood, in lymph, in chyle, in milk, and in the juices of the parenchymatous tissues. The albumin of the economy is derived from the albuminoid constituents of the food; as, casein from milk, gluten from bread, syntonin from flesh, and ovo-albumin from white of egg. These substances, though differing from each other in many respects, are by the action of the gastric juice speedily reduced and converted into peptones, in which form the distinctive characters of the different substances are lost. The peptones are absorbed into the circulation, and again undergo metamorphosis into fibrin, globulin, syntonin, casein, etc.

These substances in fulfilling their purpose in the economy break up and become oxidised; on the one hand, into certain saccharine and fatty bodies, which are ultimately eliminated in the form of carbonic acid and water; and on the other, into nitrogenous substances, which are finally eliminated by the kidneys in the form of urea.

All nitrogenous substances found in the body belong to this group of proximate principles, or are products derived from their decomposition.

From the great instability and complexity of the composition of albuminoids, their amorphous condition, and the difficulty of separating them from the inorganic substances, the phosphates, etc., that are always more or less associated with them; chemists have not been able as yet to adduce any rational formula to explain their molecular constitution. They all contain Carbon, Hydrogen, Nitrogen, and Oxygen, and most of them Sulphur and Phosphorus. The proportions of these elements vary but slightly in the different bodies; the gelatinous substances containing a rather smaller proportion of carbon, and a larger quantity of nitrogen, than the albuminous; for example,

	Albumin.	Fibrin.	Gelatin.
Carbon	53·5	52·7	50·16
Hydrogen	7·0	6·9	6·6
Nitrogen	15·5	15·4	18·3
Oxygen	22·0	23·5	24·8
Sulphur	1·6	1·2	·14
Phosphorus	0·4	0·3	

From a consideration of these numbers Lieberkühn

has given $C_{72}H_{112}N_{18}SO_{23}$, as the formula to represent the general constitution of the albuminoids.
Gerhardt believes, that the albuminoid principles are formed by the combination of an azotized
principle, of a dibasic acid character, with different saline bases in varying proportions. Mulder
maintained, that the albuminoids are formed by a
radical, which he called protein, combined with
more or less hydrogen, sulphur, and phosphorus,
according to the nature of the substance. Some
chemists have recently regarded albumin and
allied substances as nitriles of cellulose; thus
Thénard and Schützenberg state, that they obtained nitrogenous products resembling those derived from albumin, by heating such neutral hydrocarbons, as sugar, cellulose, etc., in sealed
tubes with caustic ammonia.

General characters. The albuminoids are amorphous bodies, and have never been obtained in a
crystalline form. They possess extremely low
diffusive powers. They turn the plane of polarized light to the left. Heated with caustic alkalis,
they give off ammonia, and yield leucin, tyrosin,
glycocin, and certain volatile fatty acids as formic,
acetic, and butyric acids. Distilled with sulphuric
acid and manganese peroxide, they furnish the
following products of decom position; viz., 1. All
the fatty acids from formic own to caproic acid;
2. Acetic, propionic, and butyric aldehydes; and
3. Certain aromatic acids, as collic, benzoic, and
toluic, and their nitriles. Heated with strong nitric acid they give a bright yellow colour, which

becomes orange on the addition of ammonia: this yellow colour is due to the formation of *xanthoproteic acid*. Albuminoid substances boiled with mercuric nitrate solution (Millon's test) form a red deposit, and also give a red colour to the solution. Acted upon by certain ferments, as pepsin and pancreatin, albuminoid substances become soluble and acquire greater diffusive power.

The albuminoid principles are distinguished from the gelatinous by giving precipitates with potassium ferrocyanide.

All the albuminoids are remarkable for their instability, and the readiness with which they undergo decomposition; the globulins are the least, the peptones and coagulated albumen the most stable of the whole group.

Table shewing some of the distinctive characters of the principal albuminoids.

Albumin { Soluble in water, coagulable by heat, not precipitated by acetic, or normal phosphoric acid.

Globulin { Insoluble in water, soluble in dilute acids and alkalis, precipitated by carbonic acid gas from its solutions.

Fibrin { Insoluble in water, insoluble in sodium chloride solution, soluble in solutions of potassium nitrate.

Syntonin { Insoluble in water, soluble in dilute acid solutions.

Casein {
Insoluble in water, not coaguable
by heat; is precipitated by acetic and normal phosphoric acid,
and calf's rennet.

Albuminose
or Peptones {
Soluble in water; diffusible; no
precipitate with acids; a precipitate with mercuric chloride.

ALBUMIN.

Soluble albumin, as obtained from blood serum, or white of egg, is a neutral, viscid, glairy substance, soluble in cold water, insoluble in alcohol, ether, and the essential oils. It coagulates at a temperature of 73° C. Dried at 52° C. it is converted into an amorphous transparent yellow mass, which swells up and is entirely redissolved by the addition of acetic acid. Solutions of albumin turn the plane of polarized light to the left and have a feeble diffusive power.

Albumin is coagulated by alcohol, strong mineral acids, tannic acid, carbolic acid, and by certain salts, as mercuric chloride, potassium ferrocyanide, lead acetate, and silver nitrate. Albumin is not coagulated by normal phosphoric, acetic, tartaric, or carbonic acids.

Ovo-albumin is distinguished from sero-albumin by the following characteristics. The specific rotatory power of ovo-albumin for yellow light is —35·5°; that of sero-albumin —56°. Ovo-Albumin is coagulated by ether; sero-albumin is not. Strong hydrochloric acid readily coagulates ovo-albumin;

and the coagulum is not easily redissolved; sero-albumin does not coagulate so easily, and the coagulum is more readily redissolved.

Albumin in solutions of the caustic alkalis or alkaline carbonates, is not completely coagulated by heat, a portion always remaining in solution; if however the solution be neutralized this portion is at once coagulated.

Albumin can be obtained in a tolerably pure form by beating up white of egg or blood serum with water, precipitating the other albuminoid principles with a small quantity of acetic acid, and by passing a current of carbonic acid gas through the mixture, removing the precipitated matters by filtration; the filtrate is then concentrated by a gentle heat, and submitted to dialysis.

Albumin, even after dialysis, always contains certain saline substances, which enter intimately into its composition; these substances probably are of use in keeping albumin in a state of solution in the body. The point at which the various albumins coagulate seems to depend greatly on the quantity of saline ingredients, that enter into their composition.

Metalbumin; a modification of albumin met with in dropsical fluids. It is not coaguable by heat; gives no precipitate with acetic acid or acetic acid and potassium ferrocyanide; heated with acetic acid, a slight cloudiness is given to the solution; alcohol precipitates it, but does not coagulate it.

Par-albumin. This substance was obtained by Scherer from the fluid of certain ovarian cysts; and

is usually associated with a body resembling glycogen, and which is capable of conversion into
sugar. Its solutions are extremely viscid; it is
precipitated from its warm solutions by acetic
acid, and carbonic acid gas; it also gives precipitates with lead acetate, mercuric chloride, potassium ferrocyanide and tannic acid; but it is not
precipitated by magnesium sulphate.

Coagulated Albumin. When moist, it is a white
opaque substance, which becomes yellow and
brittle when·dried; it is insoluble in water; but
soluble in strong acids and alkalis, which change
it into acid and alkali albumin respectively.

GLOBULINS.

GLOBULIN is obtained from blood serum, from
milk, from chyle, from the aqueous and vitreous
humours of the eye, and.from connective tissue;
in union with hæmatin it forms the colouring matter of the blood, "Hæmoglobin." (See Blood.)

Preparation. Rub up the fresh lenses of some
bullocks' eyes with pounded glass, agitate with
water and filter; precipitate the filtrate with carbonic acid gas.

Chemical and physical properties. Solutions of globulin become opalescent when heated to 73°; and
the globulin is deposited at 93°.

Globulin is precipitated from its solutions by
carbonic acid gas, and also by alcohol.

Ammonia does not precipitate globulin from its

solutions; but, if the ammoniated solution be neutralized with acetic acid, the globulin will be deposited. In the same manner acetic acid solutions of globulin are precipitated, when neutralized with ammonia.

Globulin in its coagulated form is insoluble in water, but, if the water be saturated with oxygen, it is dissolved; it is also soluble in neutral saline solutions.

Para-globulin. The researches of Hoppe Seyler and C. Schmidt have shown, that the globulin obtained from serum differs from that of the crystalline lens in not being precipitated from its solutions by heat or alcohol, and also by the property it possesses of coagulating certain liquids as the pericardial, peritoneal, and hydrocele fluids, which do not coagulate spontaneously themselves. This modification of globulin has been called paraglobulin, and also fibrino plastic substance from the power it has of forming with the above named fluids, fibrin. It is readily obtained by passing a stream [of carbon:·· acid gas through perfectly fresh blood serum, diluted with ten times its bulk of water; the precipitate is collected on a filter and thoroughly washed with water. If the fluid of a hydrocele be treated in the same manner, a coarse granular precipitate, very much resembling paraglobulin or fibrino plastic substance, is thrown down. This substance has been called *fibrogen* from the fact that it requires to be mixed with fibrino plastic substance to form fibrin.

MYOSIN. This substance separates and coagulates from muscle soon after death; and it is this coagulation, that most probably induces that condition of muscular rigidity, known as "rigor mortis," which after lasting a variable period gradually passes off, as the myosin becomes decomposed.

Preparation. The muscular tissue of an animal just killed is finely minced, and washed on a filter with a strong current of distilled water, till the washings give no precipitate with mercuric chloride, nor an acid reaction with test paper; the residue on the filter is then treated with a 10 per cent. solution of sodium choride and strained through linen, the resulting liquid is filtered and precipitated by the addition of distilled water; the precipitate is removed by filtration, redissolved by the sodium chloride solution, and re-precipitated with distilled water.

Chemical and physical properties. Insoluble in water, soluble in dilute acids, neutral saline, and dilute alkaline solutions. Suspended in water, it is deposited at a temperature of 70°.

Its acid solutions are not coagulated by heat; when neutralized they throw down a precipitate. Its alkaline solutions in the same manner, when neutralized with acid, are precipitated.

VITELLIN. Vitellin is most probably a mixed substance: Hoppe Seyler considers it to be a mixture of globulin and lecithin.

Preparation. The yolk of a hen's egg is agitated with ether to remove the fatty matters; and the residue treated as directed in the preparation of myosin.

Chemical characters. Vitellin closely resembles myosin, it is distinguished from it however by its greater solubility in dilute acids, and neutral saline solutions.

FIBRIN.

Fibrin in the body is always in a fluid condition, and is found in the blood, chyle, and lymph; after removal from the body it undergoes spontaneous coagulation, and forms a firm laminated clot. The formation of fibrin, as we have stated before, is due to the action of fibrinoplastic substance (paraglobulin) on fibrogen; if these two substances be mixed together in a liquid condition they combine and separate from the fluid as a jelly-like coagulum. Several circumstances may prevent or retard this coagulation; viz., extreme cold, the presence of carbonic acid, free acids as acetic, phosphoric or lactic, also alkalis and alkaline carbonates. The introduction of foreign bodies into the fluid hastens the coagulation.

Preparation. Beat up some freshly drawn blood; or better still, blood flowing from an animal, with a bundle of fine twigs; when all the fibrin is withdrawn from the blood, separate the fibrin from the twigs, place it in a muslin bag, tie the bag securely to a water-tap and allow the water to run through for some hours, from time to time removing the bag and rinsing the fibrin; continue this till the fibrin becomes free from colouring matter.

Estimation of the quantity of fibrin in blood. (See Blood.)

Chemical and physical properties. Fibrin removed from the body is a soft elastic substance of filamentous texture, which under the microscope exhibits wavy, parallel fibres of granular appearance. It is insoluble in water, cold dilute acids, alcohol, and ether; slightly soluble in cold solutions of potassium nitrate; warm solutions of dilute acid dissolve it, converting it into acid albumin. Strong mineral acids dissolve it when heated; also concentrated acetic acid, the acetic acid solution thus formed is precipitated by potassium ferrocyanide.

Fibrin decomposes hydrogen per-oxide, with the evolution of oxygen. Fibrin when dried loses about 80 per cent. of water, and is converted into a horny yellowish substance.

Lardacein. This substance is found deposited in the minute arteries of the liver, kidney, spleen, intestines, and brain, in certain diseases associated with long continued suppuration, cachexia, etc.

From the fact of its giving a blue colouration with iodine and sulphuric acid it was at first regarded as a starch and received the name of "amyloid substance." It is however a nitrogenous body, containing 15 per cent. of nitrogen and has no affinity with any kind of starch. It is regarded as a modification of fibrin; thus Virchow considers it to be an altered fibrinous deposit and the result of an oxidising action on the fibrin effused into the muscular texture of an ar-

tery; whilst Dr. Dickenson regards it as de-alka-
lized fibrin.

A comparison of the ultimate composition of
fibrin and lardacein shows how closely they are
related.

	Fibrin.	Lardacein.
Carbon	52·7	53·6
Hydrogen	6·9	7.0
Nitrogen	15·4	15·0
Oxygen	23·5	23·1
Sulphur	1·2	1·3

Lardacein is insoluble in water, dilute acids and
alkalis; treated with concentrated hydrochloric
acid it is dissolved with the formation of acid albumen; with iodine it gives a reddish brown colour,
and with iodine and sulphuric acid a blue colour,
it is not acted upon by the gastric juice.

DERIVED ALBUMINS.[o]

SYNTONIN. (Syn. natural acid albumin.) This
substance was formerly regarded as identical with
blood fibrin; it differs however from that substance in being readily soluble in dilute acids, and
in the result of its ultimate analysis, as the following table shows.

	Fibrin.	Syntonin.
C.	52·7	54·1
H.	6·9	7·2
N.	15·4	16·1
S.	1·2	1·1
O.	23·5	21·5

[o] All albuminoid substances treated with dilute acids or
alkalis, are converted into acid albumin or alkali albumin.

All albumins and globulins treated with dilute hydrochloric acid yield a substance (artificial acid albumin) which cannot be distinguished from syntonin.

Preparation. Mince muscular fibre very small, and place it on a filter; wash with a stream of cold water till the washings give no precipitate with mercuric chloride. Treat the residue on the filter with dilute hydrochloric acid (1 per cent.), and set aside for 24 hours; neutralize with sodic carbonate, and filter off the resulting precipitate which is to be well washed with cold water.

Chemical and physical properties. Syntonin is precipitated as a white opaque gelatinous mass; soluble in dilute hydrochloric acid, and in such feeble alkaline liquids as baryta and lime water. It is insoluble in water and sodium chloride solution.

Its acid solution turns the plane of polarized light to the left, is not coagulated by heat, is precipitated by a solution of magnesium sulphate, and gives a milky turbidity on the addition of sodium, calcium, or ammonium chloride.

Its alkaline solution is not coagulated by heat, and gives only a slight precipitate with magnesium sulphate.

CASEIN. (Syn. natural alkali albumin). This substance agrees in many respects with artificially formed potassium albuminate, obtained by treating albumin with a dilute solution of potassium hydrate.

Casein is a constant constituent of the milk of all mammalians; it is found also in the serum of the blood, especially in the blood of pregnent women and in the placenta. It can be obtained from the muscular plasma, and the juices of the thymus gland, and elastic tissue.

A substance resembling casein forms a large proportion of the albuminous matters of the nerve centres.

Preparation. 1. From milk. Fresh milk is precipitated by an excess of magnesium sulphate, the precipitate washed with a saturated solution of magnesium sulphate, and redissolved in water. The aqueous solution is filtered, and the filtrate precipitated with dilute acetic acid; the precipitate is separated by filtration, and washed with ether to remove the fatty matters, and dried.

2. From blood serum. A current of carbonic acid gas is passed through fresh blood serum to remove the globulin; dilute acetic acid is then added, which precipitates the casein.

Chemical and physical properties. Casein is insoluble in water, ether and alcohol; soluble in dilute acids and caustic alkalis. Its solutions are coagulated by acetic acid, and *rennet*; they are not coagulated by heat; they are precipitable by lead acetate, mercuric chloride, and cupric sulphate, with magnesium sulphate they give a precipitate soluble in water. Its acetic acid solution is precipitated by potassium ferrocyanide.

According to Hoppe Seyler, casein or natural alkali albumin differs from artificial alkali albumin in two respects. 1. Casein when treated with caustic patash yields potassium sulphide. 2. Di-

gested with artificial gastric juice, casein forms
a peptone containing phosphorus. Now artificial
alkali albumin does not yield potassium sulphide
with caustic potash, and its peptone contains no
phosphorus.   '

ALBUMINOSE or peptones; these bodies are di-
rived by the action of pepsin, in dilute acids solutions, on albuminous matters; they are extremely
diffusible and pass readily through animal membranes, turn the plane of polarized light to the left
and are not coagulated by heat, alcohol, the mineral acids, or potassium ferrocyanide. (See Digestion.)

MUCIN. This substance differs from albumin in
containing no sulphur, and in not being precipitated by heat, tannic acid, or mercuric chloride. On
the other hand, it belongs rather to the albuminoid than to the gelatinoid group, since it contains
more carbon and less nitrogen then the latter substances possess, and thus more closely approxi-
mates to the constitution of albumin, as the following table shows.

	Albumin.	Mucin.	Gelatin.
C.	53·5	52·4	50·16
H.	7·0	7·0	6·6
N.	15·5	12·8	18·3
S.	1·6	nil	·14
O.	22·0	27·8	24·8

Mucin is obtained from mucus fluids, from the
connective tissue of the embryo, from the submaxillary gland, and from tissues which have undergone mucoid and colloid degeneration.

Preparation. 1. From Mucus. Fresh Mucus is diluted with water and filtered ; the insoluble residue is treated with a weak solution of caustic potash, and filtered, the filtrate is precipitated by acetic acid, and the precipitate collected, washed well with water, acetic acid, warm alcohol, and dried.

2. From the Submaxillary gland. The submaxillary gland of an ox is rubbed down with pounded glass, and the mass placed in water for a night, then filtering and again treating the residue with water. The filtrate is precipitated by acetic acid, and the precipitate washed with water, acetic acid, warm alcohol, and then dried.

Chemical and physical properties. Mucin rapidly absorbs water swelling up, but not dissolving ; it is soluble in lime and baryta water, also in concentrated hydrochloric and nitric acid and in strong sodium carbonate solution.

Solutions of mucin are precipitated by alcohol (as a stringy mass), by acetic acid, and by the dilute mineral acids. They are not precipitated by heat, tannin, mercuric chloride, or lead acetate.

Pyin. This substance closely resembles mucin it differs however from that substance in the following respects ; the precipitate it yields with acetic acid is insoluble in excess, and it is precipitated by mercuric chloride and lead acetate. It is distinguished from chondrin by the precipitate given with alum, being insoluble in excess. Pyin can be obtained, by agitating recently drawn pus with a

10 per cent solution of sodium chloride, and allow-ing the mixture to stand for 12 hours, then filter-ing and washing the residue. The insoluble resi-due is to be treated for some hours with dilute hydrochloric acid and afterwards filtered; neu-tralize the filtrate with soda, collect the precipitate and dry.

Pyo-cyanin. A blue colour is often noticed on the dry bandages and linen which have been in contact with pus; this is due to pyo-cyanin. This substance can be obtained in a crystalline form by soaking the stained linen in water, containing a few drops of ammonia, for some hours: and then filtering off and evaporating the green liquid. The concentrated filtrate is then agitated with chlo-roform, and the chloroform solution removed and treated with very dilute sulphuric acid, and the mixture allowed to stand for some time; at length a red aqueous layer separates, which is removed and treated with caustic baryta till the solution be-comes blue; filter, and agitate the filtrate with chloroform; remove the chloroform solution and allow it to evaporate spontaneously, when pyo-cyanin will crystallize out in bluish coloured needles or rectangular flakes. The crystals are soluble in water and chloroform, insoluble in ether. Acids turn their solutions red, but alkalis restore the blue colour; chlorine decolorizes their solu-tions.

From the chloroformic solution, after the re-moval of the pyo-cyanin crystals, minute yellow crystals of pyo-xanthin can be obtained by eva-poration.

THE GELATINOUS PRINCIPLES.

These principles are distinguished from the albumins, by not being precipitated by potassium ferrocyanide, and in containing a smaller proportion of carbon, and a larger quantity of nitrogen in their composition.

GELATIN. Is obtained from bones, tendons, areolar tissue, by long boiling in water. It is sometimes found in the blood of leucocythæmic patients, and in the juice of certain carcinomatous tumours.

Preparation. From bones. Bones that have been thoroughly cleaned and dried are digested with dilute hydrochloric acid (1-20) till all the earthy matter is dissolved, the residue° is then boiled for many hours, and dried at a temperature of 100°.

Chemical and physical properties. Dry and pure gelatin is an amorphous, transparent substance, hard and brittle, with no taste, or if any feebly sweet. Insoluble in ether and alcohol. In cold water gelatin swells up without dissolving; warm water dissolves it and the solution on cooling gelatinizes, even if the solution contains only 1 per cent. of gelatin.

Solutions of gelatin are precipitated by alcohol and tannic acid; but not by potassium ferrocyanide, lead acetate, cupric sulphate, dilute mineral acids, or alum.

Solutions of gelatin are distinguished from those of chondrin by giving only a slight precipitate,

* This residue consists of a principle called *ossein* which is insoluble in water : by long boiling it is converted into gelatin

soluble in excess, with acetic acid; and no precipitate with solution of alum.

Boiled with sulphuric acid it yields leucin and glycocin.

CHONDRIN. Is obtained from cartilaginous tissue. Young bones prior to ossification, and adults bones in certain diseased conditions, yield considerable quantities.

Preparation. The costal cartilages of the calf are cut in thin slices, boiled for 24 hours, and the solution evaporated to a gelatinous consistence; the fatty matters are removed by digestion with boiling ether, and dried at a temperature of 100°.

Chemical and physical properties. Chondrin is a diaphanous, horny substance, insoluble in alcohol, and ether; on the addition of cold water it swells up (to about 12 times its original bulk) but is not dissolved; it dissolves freely in boiling water. It gives only a slight precipitate with tannic acid; it also gives precipitates with dilute mineral acids, and metallic salts, but not with potassium ferrocyanide. It gives a precipitate, insoluble in excess, with acetic acid; and a precipitate, soluble in excess with solution of alum.

Boiled with sulphuric acid it yields leucin but no glycocin.

ELASTICIN. This substance is the special principle of yellow elastic tissue, and is consequently obtained from those textures in which this tissue is most · abundant; as the yellow elastic ligaments of the vertebræ, the ligamentum nuchæ, the middle coat of arteries and veins, the arcolar tissue, and the lower vocal cords.

Preparation. The ligumentun nuchæ, or the middle coat of the arteries or veins, are boiled with alcohol and ether to remove the fatty matter; then treated with water at a temperature of 100° for 24 hours; and afterwards for 1 hour with water at 120°, this removes the other gelatinous principles; the residue is then boiled with strong acetic acid, and washed with water; again boiled with strong soda ley and treated with strong acetic acid, and washed with water; finally the residue is digested with hydrochloric acid, washed with water, and dried.

Chemical and physical properties. Elasticin forms a yellow, fibrous, brittle mass, soluble only in strong caustic alkalis; it it insoluble in water below the temperature of 120°. Burnt on platinum foil it leaves no residue. . Elasticin is quite free from sulphur.

KERATIN. Is obtained by treating pounded horny matter, as epithelium, epidermis, feathers, hoofs, horns, nails etc., with boiling alcohol and ether; the residue, which however is far from pure, is Keratin.

Chemical and physical properties. Keratin by long boiling is dissolved, but the solution on cooling does not gelatinize. Acetic acid causes it to swell up but does not dissolve it. Caustic potash dissolves it, also strong sulphuric acid; the acid solution is rendered turbid by potassium ferrocyanide.

PART II.

PRODUCTS OF DECOMPOSITION.

WE have already stated that the highly complex organic substances entering into the composition of the animal tissues and fluids are decomposed in the body, by the action of oxygen introduced into the economy by the process of respiration, into less complex molecules.

These products of decomposition are divided into two distinct groups, viz.: 1. The non-nitrogenous organic acids derived directly from the oxidation of the saccharine and oleaginous principles, and indirectly from the albuminous; and, 2. The nitrogenous bases or *amides*, which are obtained, together with the non-nitrogenous organic acids, from the oxidation of albuminous and gelatinous principles.

CHAPTER IV.

THE NON-NITROGENOUS ORGANIC ACIDS.

THESE acids may be distributed into two series; 1. The fatty acid series; and, 2. The aromatic acid series.

FATTY ACIDS.

These acids are derived, by the oxidation of their corresponding alcohols, from the "homologous series of hydro-carbon radicals." They are arranged in groups according as they are formed by monatomic or diatomic radicals.

Group I.—Monatomic Fatty Acids.

NAME.	FORMULA	BOILING POINT.	NAME.	FORMULA.	BOILING POINT.	
Formic	CH_2O_2	100°	Caproic	$C_6H_{12}O_2$	199°	
Acetic	$C_2H_4O_2$	118°	Capric	$C_8H_{16}O_2$	236°	
Propionic	$C_3H_6O_2$	140°	Palmitic	$C_{16}H_{32}O_2$	62°	Melting pt.
Butyric	$C_4H_8O_2$	162°	Stearic	$C_{18}H_{36}O_2$	69°	
Valeric	$C_5H_{10}O_2$	174°	Oleic*	$C_{18}H_{34}O_2$	14°	

The acids enumerated in the above table (with the exception of oleic acid) are derived from the " monatomic series of homologous hydro-carbons" by the oxidation of the corresponding alcohols, in which 1 atom of oxygen replaces 2 atoms of hydrogen; thus, Ethyl alcohol by oxidation loses 2 atoms of hydrogen and is converted into aldehyde; and aldehyde by farther oxidation bebecomes acetic acid.

Ethyl alcohol.   Aldehyde.

$$C_2H_6O \quad + O \quad = \quad C_2H_4O \quad - H_2O$$

Aldehyde.   Acetic Acid.

$$C_2H_4O \quad + O \quad = \quad C_2H_4O_2$$

* Belongs to the Glycerin Series.

These acids are volatile and liquid at temperatures above 17° (with the exception of Palmitic and Stearic acid which are solid); they redden litmus paper; are monobasic, and unite with bases to form salts more or less soluble in water, the solubility of the salt diminishing proportionately for every additional atom of CH_2 present in the acid; their solutions when shaken give a persistent lather. The boiling point of these acids increases for every additional atom of CH_2; thus, Formic acid CH_2O_2 boils at 100° C; Acetic acid $C_2H_4O_2$ at 118° C; Propionic acid $C_3H_6O_2$ at 140°, and so on.

Formic, Acetic, and Propionic acids are present in sweat, but in no other healthy human secretion. In diseased states, when there is diminished oxidation, they have been found in the blood, the muscular plasma, and the splenic juice. In certain forms of dyspepsia they have been obtained from the vomited matters; and are present in stale, or diabetic, urine in which acid fermentation has commenced.

Butyric acid is met with in human sweat, especially the sweat of the external genitals; it is also formed in considerable quantities in the first part of the large intestine; with capric, oleic, stearic acid, etc., in union with glycerin, it is found in milk. As an abnormal product, it is sometimes met with in the blood, muscular juice, in the fæces, and in the urine.

Capric and caproic acids in union with glycerin (as glycerides) are found in the sweat and milk. Valeric acid is obtained from the juice of flesh, and Frerichs found it associated with leucin in the urine in a case of typhus fever.

Palmitic, stearic and oleic acids with glycerin (glycerides) form the neutral animal fats (see p. 13.)

The various acids are distinguished from each other, by their physical characters, such as consistence, the difference in their boiling points; by the various degrees of solubility of their salts, and the characteristic odour many of them evolve.

FORMIC ACID. CH_2O_2.

Is a colourless corrosive liquid, boiling point 100°, solid at 1°.

ACETIC ACID. $C_2H_4O_2$.

Is a colourless sharp smelling liquid, boiling point 118°, solid at 17°.

PROPIONIC ACID. $C_3H_6O_2$.

Is a colourless oily liquid boiling point 140°, solid at 20°.

BUTYRIC ACID. $C_4H_8O_2$.

Is a colourless mobile liquid, boiling point 162°, solid at 20°, odour of rancid butter.

VALERIC ACID. $C_5H_{10}O_2$.

Is a limpid colourless oily liquid, boiling point 174°, solid at 20°, odour of valerian.

Caproic Acid. $C_6H_{12}O_2$.

An oily liquid having the odour of acid sweat, boiling point 199°, solid at 4°.

Capric Acid. $C_8H_{16}O_2$.

Is a greasy oily liquid, which crystallizes at 29° in colourless needles which on heating evolve a goaty odour, boiling point 236°.

Palmitic Acid. $C_{16}H_{32}O_2$.

Is a tasteless, white, fatty substance, melting point 62°, soluble in ether and alcohol forming acid solutions which on concentration deposit white crystallin needles; insoluble in water. With glycerin it forms three bases (glycerides), 1. Monopalmitin, 2. Dipalmitin, and 3. Tripalmitin; the latter is a constituent of animal fat.

Stearic Acid. $C_{18}H_{36}O_2$.

Is a white crystaline substance, melting point 69·2° soluble in ether and alcohol, insoluble in water. Like palmitic acid it forms with glycerin three compounds, Monostearin, Distearin and Tristearin, the latter is a constituent of suet.

Oleic Acid. $C_{18}H_{34}O_2$.

Is solid at 4°, liquid at 14°; freely soluble in alcohol and ether, insoluble in water. By the action of nitrous acid it is converted into elaidic acid. By distillation it yields sebacic acid, this distinguishes it from other fatty acids. Heated with caustic potash it loses 2 atoms of hydrogen and is converted into acetic and palmitic acids. With glycerin it forms Monolein, Diolein, and Triolein, the latter forms the oily portion of animal fat.

Oleic acid is derived from the glycerin series of ." homologous hydro-carbons."

Separation of the Volatile Fatty acids from Organic Mixtures. Introduce the mixture into a strong wide necked flask fitted with a tube containing a thermometer, and communicating with a condenser; a few drops of sulphuric acid are added to the mixture and heat applied. The more volatile acids pass over first and are received into a beaker, the less volatile collecting in the wide neck of the flask condense and fall back into the mixture. When the distillate comes over, the point at which the thermometer stands is observed, and the beaker removed when a certain temperature is reached; another beaker is then placed to receive the next portion of the distillate, and so on. Finally the different beakers are in turn submitted to redistillation till a constant boiling point is obtained, which corresponds to that of the acid supposed to be present.

Separation of Palmitic, Stearic and Oleic acids from Organic Mixtures. Exhaust the mixture thoroughly with boiling alcohol, and filter; boil the filtrate with solution of potash which saponifies the fatty matters. The Potassium Palmitate, Stearate and Oleate are then removed, and decomposed by the

E

addition of alcohol containing a small quantity of sulphuric acid, the precipitated potassium sulphate being removed by filtration; the alcoholic solution on concentration will deposit Palmitic and Stearic acid in a crystalline form ; after these have been separated the alcoholic solution is evaporated, and extracted with ether, and the etherial solution distilled off leaving oleic acid as a residue.

GROUP II.—DIATOMIC FATTY ACIDS.

MONOBASIC.		DIBASIC.	
CARBONIC HYDRATE	CH_2O_3	OXALIC	$C_2H_2O_4$
GLYCOLLIC . . .	$C_2H_4O_3$	SUCCINIC	$C_4H_6O_4$
LACTIC	$C_3H_6O_3$	SEBACIC	$C_{10}H_{18}O_4$
LEUCIC	$C_6H_{12}O_3$		

These acids are derived from the "olefine series of homologous hydrocarbons" by the oxidation of the corresponding alcohols, and may be divided into two classes; viz., the *monobasic* acids which are formed by one atom of oxygen replacing two atoms of hydrogen in the corresponding alcohol; and the *dibasic* acids which are formed by the replacement of four atoms of hydrogen by two atoms of oxygen; thus, ethylene alcohol by oxidation loses two atoms of hydrogen, and is converted into monobasic, glycollic acid; and glycollic acid by further oxidation loses two atoms of hydrogen and becomes dibasic, oxalic acid.

Ethylene Alcohol.   Glycollic Acid.

$$C_2H_6O_2 \ + \ O_2 \ == \ H_2O \ + \ C_2H_4O_3$$

Glycollic Acid.   Oxalic Acid.

$$C_2H_4O_3 \ + \ O_2 \ = \ H_2O \ + \ C_2H_2O_4$$

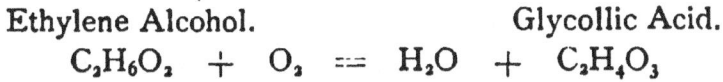

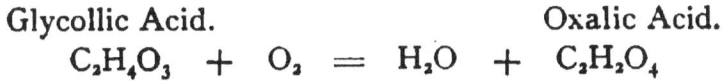

MONOBASIC ACIDS.

CARBONIC ACID. (See Carbonates. Part III.)

GLYCOLLIC ACID. $C_2H_4O_3$.

This substance does not exist in a free state in the organism. Its ammoniated form, glycocin, conjugated with cholic acid forms the glycocholic acid of the bile; and with benzoic acid unites to form hippuric acid. Glycollic acid is a syrupy, acid liquid, soluble in ether and alcohol, from concentrated solutions of which, deliquescent crystals which melt at 78° are deposited. It may be prepared by treating glycocin with nitrous acid; thus,

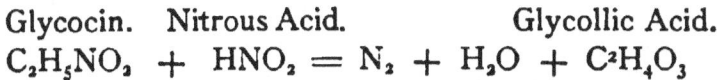

Glycocin.   Nitrous Acid.   Glycollic Acid.

$$C_2H_5NO_2 \ + \ HNO_2 \ = \ N_2 \ + \ H_2O \ + \ C^2H_4O_3$$

LACTIC ACID. $C_3H_6O_3$.

Lactic acid in a free state is formed in the muscular plasma, and gives to that fluid its acid reaction, the quantity present is increased by muscular contraction. Associated with hydrochloric acid it is almost invariably met with in the gastric juice. It is never obtained from healthy blood, since in that fluid its salts are speedily de-

E 2

composed and converted into alkaline carbonates. In diseases, however, in which the oxidating power is diminished, as in some forms of dyspepsia, pyrexia, and pulmonary affections, lactic acid may be found in the blood and urine. When present in the urine it is generally associated with an excess of calcium oxalate, another evidence of defective oxidation. In rachitic children the urine contains lactic acid, associated with an abundance of calcium phosphate, hence it has been suggested that an excess of lactic acid in the blood holds the calcareous salts in solution, and consequently they pass out of the system in the urine instead of being deposited and forming bone. Dr. Richardson considers that the presence of free lactic acid in the blood produces the phenomenon of Rheumatism, as he found the injection of lactic acid into the peritoneal cavity of a dog induced peritonitis and endocarditis. Lactic acid is formed in the system by the oxidation of the saccharine and albuminous principles. It plays an important part in the performance of the functions of the body; 1, by its power of holding calcareous salts in solution it prevents the deposition of bony matter which would otherwise accumulate in all the tissues; 2, by the rapid combustion of its salts into alkaline carbonates it furnishes a supply of heat to the economy; and 3, by its presence in the intestinal canal during digestion it promotes the absorption of food into the blood-vessels and lacteals.

To obtain lactic acid from an organic mixture. The fluid is gently evaporated, at a temperature a lit-

tle below 100° C., to one fifth its bulk, and then filtered; to the filtrate baryta is added and the precipitate removed by filtration; to the filtrate add a
few drops of strong sulphuric acid and the mixture
gently distilled; the residue left after distillation
is then to be shaken with alcohol and allowed to
digest. After standing some days it is filtered and
the filtrate mixed with milk of lime and evaporated to dryness; the residue is dissolved in water
and a stream of carbonic acid gas passed through
the solution, which is to be heated to 100° C.
When the solution is cold the precipitate is removed by filtration; the filtrate evaporated to
dryness, the residue dissolved in rectified alcohol, and the alcoholic solution concentrated and
set aside; in a few days characteristic crystals of
calcium lactate will be deposited.

Chemical and physical properties. Lactic acid is a
colourless, syrupy fluid of sharp acid taste; specific
gravity 1·21; soluble in water, alcohol, and ether.
If distilled at temperatures above 160° it decomposes. Heated with sulphuric acid it evolves
·carbonic oxide. Heated with nitric acid it yields
oxalic acid. Its calcium and zinc salts are characteristic. *Sarcolactic acid* closely resembles lactic
acid and is isomeric with it; its salts however differ
in crystallizing with a smaller proportion of water,
and in their crystalline form, it is obtained from
muscular tissue.

LEUCIC ACID. $C_6H_{12}O_3$.

This acid only exists in the body in its ammoniated form, leucin, from which it can be obtained by heating with nitrous acid. (See leucin).

DIBASIC ACIDS.—OXALIC ACID. $C_2H_2O_4$.

Oxalic acid represents the last intermediate stage in the oxidation of the more complex organic substances into carbonic acid and water. In a state of health the oxalates are never met with in any of the fluids and secretions of the body, as they at once undergo combustion, and are converted into carbonic acid and water; but if the process of oxidation be impeded then oxalates will be met with in the urine. We may therefore expect to find oxalates in urine in all cases of debility where the oxidizing power is diminished, as in dyspepsia, or in the convalescence after severe fevers, or in chronic diseases of the respiratory organs, as in chronic bronchitis, and emphysema. Oxalates are also formed in the urine after drinking carbonated beverages as champagne, seltzer water etc., and after eating certain fruits and vegetables, as rhubarb, sorrel, etc., which contain them in large quantities.

Oxalic acid crystallizes in large, four-sided, transparent prisms, which contain two molecules of water of crystallization; the crystals are soluble in 9 times their weight of cold water, they are slightly soluble in alcohol, their solutions have an intensely sour taste. Heated to 160° oxalic acid is partially decomposed into formic acid, carbonic anhydride, and water. With solutions of lime it forms the normal calcium oxalate CaC_2O_4, a salt of great interest to the pathological chemist. (See calcium oxalate, Urine).

SUCCINIC ACID. $C_4H_6O_4$.

This acid has been obtained from certain morbid exudations, as the contents of hydatid cysts, hydrocele, etc., by evaporating the fluids to the consistence of syrup and adding hydrochloric acid and thoroughly exhausting the acid solution with ether, the etherial solution, on distillation, yields crystals of succinic acid. The crystals form large, rhombic, colourless tablets which fuse at 160°, and are soluble in alcohol and cold water; they are not decomposed by a high temperature, and can therefore be removed from organic mixtures by destructive distillation. Solutions containing succinic acid give a brown precipitate with ferric chloride, and white with barium chloride.

AROMATIC ACID SERIES.

BENZOIC ACID. $C_7H_6O^2$
PHENOL. C_6H_6O

Benzoic acid $C_7H_6O_2$ is formed in the urine of all herbivorous animals and can be obtained from stale human urine; it does not exist in a free state in any of the tissues and fluids of the human body, and its chief interest is due to the presence of its radical in hippuric acid. Benzoic acid occurs in pearly-white crystalline plates which fuse at 121°. Benzoic acid and its salts give reddish-brown precipitates with ferric chloride, and blue precipitates with cupric acetate.

Phenol C_6H_6O (syn. carbolic acid or phenol alcohol).* This substance associated with *taurylic*, *damaluric*, and *damolic acids* is met with in extremely minute quantities in human urine; the quantity is increased by the internal administration of carbolic acid, kreasote, coal tar, etc. The process of obtaining and separating these acids from the urine is long and tedious, and as they possess little practical interest for the ordinary student it is not described here.†

The following is a brief description of the chief characters of these four acids. Phenol occurs as a white crystalline mass melting at 42° and forming a heavy, oily, corrosive fluid with a pungent, smoky odour. It gives a violet colour to solutions of ferric chloride, which acquires a blue colour on exposure to the air; a chip of fir wood saturated with phenylic acid and dipped into dilute hydrochloric acid turns a deep blue colour.

Taurylic acid is a colourless, oily liquid, fluid at 18°; mixed with an equal volume of concentrated sulphuric acid it is converted into a solid mass, the mother liquor of which contains phenol

---

* The phenols of which carbolic acid is the type appear to be intermediate in properties between alcohols and acids.

† See Neubauer and Vogel *On the Urine*, p. 41.

sulphuric acid. Damaluric acid is an oily fluid having an odour of valerian, it is distinguished from damolic acid by the infusibility of its baryta salt.

THE RESINOUS ACIDS.

CHOLESTERIC ACID	$C_8H_{10}O_5$
CHOLIC ACID	$C_{24}H_{40}O_5$
LITHOFELLIC ACID.	$C_{20}H_{38}O_4$

These acids and cholesterin are probably derived from the same radical; since on oxidation they yield the same products of decomposition. The radical, however, has not yet been isolated, but it is probable that it belongs to some of the higher aromatic hydrocarbons since cinnyl alcohol $C_8H_{10}O$ is homologous with cholesterin.

CHOLESTERIC ACID. $C_8H_{10}O_5$.

This acid is formed, together with acetic and other volatile acids, whenever cholesterin, cholic, or lithofellic acids are heated with nitric acid. It is a yellow, non-crystallizable substance, which rapidly absorbs moisture from the air, it is soluble in water, alcohol, and ether, the solutions have an acid and bitter taste.

CHOLIC ACID. $C_{24}H_{40}O_5$.

Is found in bile combined with glycocin and taurin, forming with these substances the well known bile acids. When bile has been removed

some time from the body, the bile acids undergo
decomposition and set free cholic acid.°

Preparation. Purified ox bile is boiled with barium hydrate in an apparatus arranged so as to allow the aqueous vapours that arise, to condense and
fall back again; after boiling about 12 hours the
mixture is set aside to cool, and when cold is su-
persaturated with hydrochloric acid. The resinous
precipitate thus obtained is washed and redissolved
in soda and again precipitated with the acid; the
precipitate is now allowed to stand for some time
in contact with ether, which changes it into a crystalline mass, this is dissolved in hot alcohol, and
distilled water is then added till the mixture becomes turbid; after standing some time crystals of
cholic acid are deposited.

Chemical and physical properties. Cholic acid occurs in two forms, the amorphous and the crystalline. The former is resinous and viscous, very
slightly soluble in water, but freely soluble in
alcohol and caustic alkalis. In the latter the
crystals are octohedral and tetrahedral, they are
colourless, insoluble in water, very soluble in ether
and alcohol; the octahedral variety contains 1
molecule, the tetrahedral $2\frac{1}{2}$ molecules of water;
when heated they lose this water of crystallization
and become disintegrated. Cholic acid heated with
acids at a temperature of 200° C. loses 2 atoms of
water and is converted into dyslysin (so named from
its insolubility in water, acids, alkalis, and alcohol.

* Pigs bile contains hyo-cholic acid $C_{25}H_{40}O_4$ conjugated
with glycocin and taurin.

Cholic Acid.   Dyslysin.

$$C_{24}H_{40}O_5 \; - \; 2H_2O \; = \; C_{24}H_{36}O_3$$

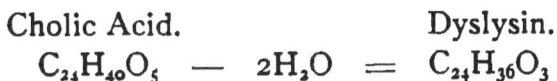

Treated with nitric acid it is decomposed, cholesteric acid being one of the products. Cholic acid with a solution of sulphuric acid and sugar, gives a deep purple coloration to the mixture; this reaction is characteristic of the bile acids and is known as Pettenkoffer's test.

LITHOFELLIC ACID $C_{20}H_{38}O_4$. This substance is obtained from certain concretions known as bezoars, by treating them with boiling alcohol; on cooling, the alcoholic solution deposits impure crystals of the acid; these are purified by washing with cold alcohol and redissolving in boiling alcohol, and filtering the solution through animal charcoal. Lithofellic acid crystallizes in minute rhombic prisms, with an oblique terminal face. The crystals are insoluble in water, less soluble in alcohol and insoluble in ether; they melt at a temperature of 204°, but on cooling they again form a crystalline mass; if heated beyond their melting point the mass on cooling is resinous. Submitted to dry distillation lithofellic acid loses 1 atom of water and is converted into pyrofellic acid. Burnt in air, lithofellic acid gives off white vapours having an aromatic odour. Boiling hydrochloric acid converts it into a brown resinous substance resembling cholic acid. With Pettenkoffer's test lithofellic acid gives the same purple coloration as observed with cholic acid.

CHAPTER V.

THE ANIMAL NITROGENOUS BASES.

THESE bodies are formed by the oxidation of the albuminous and gelatinous principles, and are called Amides. They may be considered to consist of ammonia, in which one or more atoms of hydrogen are replaced by an acid organic radical; and they may be conveniently arranged in three groups.

1. The *Monamides*, which are formed by a single molecule of ammonia $\left. \begin{array}{l} H \\ H \\ H \end{array} \right\} N.$ in which one or more atoms are replaced by an acid organic radical; for example, in the case of hippuric acid $C_9H_9NO_3$, 1 atom of benzoic acid radical C_7H_5O and 1 atom of glycollic acid radical $C_2H_3O_2$ replace 2 atoms of H from the typical ammonia molecule $\left. \begin{array}{l} H \\ H \\ H \end{array} \right\} N.$ to make $\left. \begin{array}{l} C_2H_3O_2 \\ C_7H_5O \\ H \end{array} \right\} N.$ and this view of the constitution of hippuric acid is expressed by the term benzamido glycollic acid.

2. The Primary *Diamides*, or Ureas, are formed by a double molecule of ammonia $\left.\begin{array}{c} H_2 \\ H_2 \\ H_2 \end{array}\right\} N_2$ in which two or more atoms of H are replaced by an organic acid radical; thus, we have urea, in which the dibasic radical of carbonic acid CO″ replaces 2 atoms of H in the double molecule of ammonia $\left.\begin{array}{c} H_2 \\ H_2 \\ H_2 \end{array}\right\} N_2$, to form $\left.\begin{array}{c} CO″ \\ H_2 \\ H_2 \end{array}\right\} N_2$ or in other words a diamide of carbonic acid.

3. The Secondary *Diamides*, or Ureides, are closely related to the preceding group; they contain either one or two residues of urea respectively, (monureides and diureides,) and a residue of a simple non-nitrogenous acid, as carbonic, oxalic, or mesoxalic acid; for example, uric acid when oxidized in the presence of water loses two atoms of hydrogen and is converted into de-hyduric acid; thus,

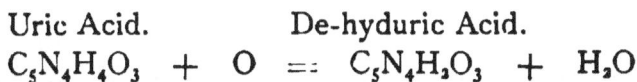

Uric Acid. De-hyduric Acid.

$$C_5N_4H_4O_3 \; + \; O \; =\!\!= \; C_5N_4H_2O_3 \; + \; H_2O$$

and de-hyduric acid by further oxidation is converted into alloxan and urea.

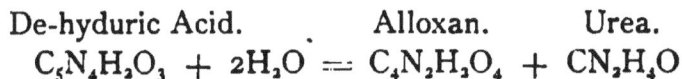

De-hyduric Acid. Alloxan. Urea.

$$C_5N_4H_2O_3 \; + \; 2H_2O \; =\!\!= \; C_4N_2H_2O_4 \; + \; CN_2H_4O$$

and alloxan furnishes mesoxalic acid and another atom of urea,

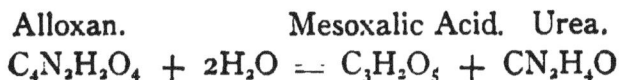

Alloxan. Mesoxalic Acid. Urea.

$$C_4N_2H_2O_4 \; + \; 2H_2O \; =\!\!= \; C_3H_2O_5 \; + \; CN_2H_4O$$

showing that de-hyduric acid contains two residues of urea and one residue of mesoxalic acid.

GROUP I.   MONAMIDES.

GLYCOCIN	$C_2H_5NO_2$	CYSTIN	$C_3H_7NSO_2$
LEUCIN	$C_6H_{13}NO_2$	HIPPURIC ACID	$C_9H_9NO_3$
SARCOSIN	$C_3H_7NO_2$	TYROSIN	$C_9H_{11}NO_3$
CHOLIN	$C_5H_{15}NO_2$	CEREBRIN	$C_{17}H_{33}NO_3$
TAURIN	$C_2H_7NSO_3$	LECITHIN	$C_{42}H_{84}PNO_9$

GLYCOCIN. $C_2H_5NO_2$.

Is derived from the decomposition of the gelatinous principles of the body; it does not exist in a free state in the economy, but conjugated with cholic acid it forms glycocholic acid.

Glycocin is formed by the replacement of one atom of H in the molecule of ammonia by the monad radical of glycollic acid $C_2H_3O_2$; thus

$$\left.\begin{array}{c} C_2H_3O_2 \\ H \\ H \end{array}\right\} N.$$

Preparation. Glycocholic acid is boiled with an excess of baryta water, the precipitate of barium cholate removed, and the solution concentrated; after standing a little time glycocin will crystallize out. Glycocin is also obtained by boiling gelatin for some time with dilute acids.

Chemical and physical properties. Crystals of glycocin are hard and granular, and have a sweet, mawkish taste; they are soluble in 400 parts of cold water, but quite insoluble in alcohol. A transient fiery red colour is given when glycocin is heated with a strong solution of caustic potash. A stream of nitrous acid passed through an aqueous solution of glycocin decomposes it, nitrogen is evolved, and on agitating the solution with ether, and evaporating, glycollic acid is obtained.

LEUCIN. $C_6H_{13}NO_2$.

This substance, associated with tyrosin, may be obtained from all glandular organs and their secretions; it is especially abundant in the lung and liver tissue. It is never found in normal urine, its presence in that fluid always denoting serious disease; thus, it has been met with in the urine, in severe cases of jaundice, in acute yellow atrophy of the liver, in cirrhosis of that organ, and in cases of smallpox and typhus, in these diseases the quantity is also increased in the other organs of the body.

Leucin may be regarded as formed from ammonia by the replacement of 1 atom of H by the leucic acid radical $C_6H_{11}O_2$; thus,

$$\left.\begin{array}{l} C_6H_{11}O_2 \\ H \\ H \end{array}\right\}N.$$

Preparation. Leucin can be obtained by fusing albumin, casein, gelatin, yellow elastic tissue, or epidermal appendages, with an equal weight of potassium hydrate, after the reaction set up has continued some time, and the mass turns yellow; it must be dissolved in hot water, and this solution contains leucin and tyrosin; the latter is precipitated by acetic acid and removed by filtration. The mother liquor is now concentrated, and on cooling leucin is deposited in crystalline plates, which must be re-crystallized from alcohol to obtain them pure.

When present in urine, it is only required to evaporate about 5 ounces of that fluid to a thin syrup and when cold, leucin in the shape of oily circular-looking discs will be deposited.

Fig 1..
Leucin in oily discs.

Chemical and physical properties. Leucin is deposited from its alcoholic solution in white shining plates, greasy to the touch, lighter than water, and much resembling cholesterin in appearance, it is distinguished from that substance by its insolubility in ether. It is slightly soluble in cold water, and very soluble in boiling water, soluble in 600 parts of cold absolute alcohol, very insoluble in ether, melting point 170°; is decomposed by nitrous acid, leucic acid being formed and nitrogen given off. Distilled with dilute sulphuric acid and manganese peroxide, it yields valero-nitrile, carbonic acid, and water. Fused with caustic potash it is transformed into potassium valerate, hydrogen and ammonia.

Leucin obtained from urine is not crystalline, but forms circular oily-looking discs which float on the surface of water; they generally have a somewhat yellowish appearance owing to the colouring matter of the urine; if this form of leucin be dissolved in boiling alcohol the solution on cooling will deposit leucin in crystalline plates.

SARCOSIN. $C_3H_7NO_2$.

This substance is only interesting to the physiological chemist from being a constituent of kreatin ; it has never been directly obtained from any of the solids or fluids of the body.

Sarcosin is formed by one atom of glycollic acid radical $C_2H_3O_2$, and one atom of methyl CH_3 replacing two atoms of H from the molecule of ammonia ; thus,

$$\left.\begin{array}{l} C_2H_3O_2 \\ CH_3 \\ H \end{array}\right\}N;$$ Sarcosin is therefore considered as Methyl glycocin.

Preparation. Kreatin is boiled with baryta water till no more ammonia is evolved, the excess of baryta is then got rid of by passing a stream of carbonic acid gas through the [mixture, which is then boiled for some time, and the barium carbonate removed by filtration. The filtrate is treated with an excess of dilute sulphuric acid, and evaporated ; the residue is then dissolved in water to which a little barium carbonate has been added, and the mixture boiled, and filtered, and the filtrate evaporated to a thin syrup, from which the Sarkosin will crystallize out.

Chemical and physical properties. The crystals are right rhombic prisms which melt at 100°, and sublime without leaving any residue, they are colourless, very soluble in water, insoluble in ether. An aqueous solution of sarkosin gives with platinic chloride a yellow double salt, and with cupric acetate a blue double salt.

CHOLIN. $C_5H_{15}NO_2$.   (*Syn.* neurin).

This energetic base was obtained originally by Strecker in 1861 from pigs' bile, it is also a constituent of lecithin, protagon, and cerebrin. It is apparently one of the products of the metamorphosis of nervous tissue.

Cholin may be regarded as ammonium hydrate, in which 1 atom of H is replaced by ox-ethyl C_2H_5O + 3 atoms of methyl CH_3 making

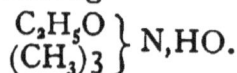

$$\left. \begin{array}{c} C_2H_5O \\ (CH_3)3 \end{array} \right\} N,HO.$$

Preparation. The alcoholic solution of bile (see Bile) is precipitated with ether, and the etherial solution evaporated; the residue boiled for some hours with baryta water, and the excess of baryta removed by precipitation with carbonic acid. The filtrate is concentrated, mixed with absolute alcohol and filtered; and the filtrate neutralized with hydrochloric acid, and set aside till the taurin separates out, which is removed by filtration. To the filtrate some solution of platinic chloride is added when a yellow crystalline precipitate of cholin platino chloride is thrown down; this is separated by filtration, dried, and treated with moist silver oxide, the cholin then separates as a syrupy liquid.

Cholin can be obtained from the etherial solution of nervous tissue by the same process.

Taurin. $C_2H_7NSO_3$.

This substance associated with cholic acid is a constant constituent of the bile, forming taurocholic acid; it can be obtained from all glandular tissues, especially from the lungs, and from voluntary muscular fibre. From which of the albuminoid constituents it is. derived and the manner in which it is separated, we are still ignorant.

Taurin or ethyl amido sulphuric acid, or properly ethyl sulphanate, is formed by 1 atom of ethyl C_2H_5 and 1 atom of sulphuric acid radical SO_2HO replacing 2 atoms of H in the molecule of ammonia; thus, $SO_2HO \left.\begin{array}{c} C_2H_5 \\ \\ H \end{array}\right\} N.$

Preparation. The alcoholic extract of bile is boiled for some hours with hydrochloric acid which precipitates the resinous matters, these are filtered off, and the filtrate evaporated till nearly dry; the crystals of sodium chloride which are deposited being removed, boiling alcohol is then added and on cooling the solution deposits impure crystals of taurin, these must be washed with water and recrystallized.

Chemical and physical properties. The crystals are four-sided prisms with pyramidal extremities, insoluble in ether and alcohol, soluble in 15 parts of cold water; the aqueous solution has a neutral reaction. They are dissolved by the mineral acids without change; burned in air they evolve sulphurous acid fumes; heated with potash, ammonia is evolved, and potassium sulphate formed. The aqueous solutions are not precipitated by salts of mercury, copper, or silver.

CYSTIN. $C_3H_7NSO_2$.

According to Dr. Bence Jones cystin is constantly being separated in the healthy organism, immediately undergoing transformation into sulphuric acid, carbonic acid, and urea. Whenever this chemical transformation is arrested cystin appears in the urine. As the composition of cystin is $C_3H_7NSO_2$ the proportion of nitrogen to carbon is four to twelve, in uric acid $C_5N_4H_4O$ it is four to five, and in urea CN_2H_4O four to two; therefore 12, 5, and 2 are the indices representing the different amounts of of suboxidation in cystin, uric acid, and urea respectively. Dr. Bence Jones thus regards cystin as representing the smallest degree of the oxidation of the albuminous principles, in the same way that sugar in diabetes represents the least degree of oxidation of the amylaceous principles.

Cystin calculi which are rare, are soft, compressible, and often have a waxy appearance externally; they are only formed in urine which is neither very acid nor alkaline.

Preparation. The urine supposed to contain cystin is acidulated with acetic acid and allowed to stand till a deposit is formed. This deposit is dried, redissolved in ammonia, and the ammoniacal solution evaporated, cystin if present will be recognized in the residue by the hexagonal form of its crystals.

Chemical and physical properties. The crystals are hexagonal laminæ, which overlay each other, forming little groups, they have a pale lemon colour which on exposure to light and air acquires a

Fig. 2. Cystin. greenish tinge. They are soluble in strong nitric acid and the acid solution evaporated with ammonia, does not give the murexide reaction (distinguishing them from uric acid crystals). Caustic alkalis also dissolve cystin, the alkaline carbonates precipitate it from its acid solutions, and acetic acid from its alkaline solution. Heated in air the crystals give off vapours of hydrocyanic acid; by dry distillation they yield fumes of ammonia, hydrocyanic acid, and an insoluble residue; treated with silver oxide they yield silver sulphide, ammonia sulphate, and a yellow amorphous residue very soluble in water.

HIPPURIC ACID. $C_9H_9NO_3$.

This substance is a normal constituent of human urine, the quantity passed in the twenty-four hours under ordinary circumstances varying from 0·8 to 1 grm. Weisman however gives 2·17 grms. as the normal daily quantity excreted.

The excretion is greatly augmented by a vegetable diet, and especially by such vegetable substances as benzoic acid, cranberries, blackberries, and plums. Consequently we are not surprised to

find a considerable quantity in the urine of all herbivorous animals; thus cow's urine contains 1 per
cent., and horse's urine 0·38.   In these animals
hippuric acid often undergoes oxidation in the system, and is converted into benzoic acid which
appears in the urine; thus horses at rest pass
urine free from benzoic acid and containing the
standard quantity of hippuric acid, but when put to
hard work the hippuric acid diminishes and benzoic acid appears.

Kühne has observed that benzoic acid given to
patients suffering from disease of the liver passes
unchanged into the urine instead of being converted into hippuric acid, which would have been
the case under ordinary circumstances. From
this fact he has assumed that hippuric acid is derived from the vegetable aromatic constituents of
our food and the place of their transformation is
the liver.

The excretion of hippuric acid is increased in all
febrile affections, also in diabetes.

Hippuric acid contains residues of benzoic acid
(C_7H_5O) and glycollic acid ($C_2H_3O_2$) which replace 2 atoms of hydrogen from a molecule of am-

monia; $\left. \begin{array}{l} C_7H_5O \\ C_2H_3O_2 \\ H \end{array} \right\} N$ and this constitution is ex-

pressed by the term benzamido glycollic acid.

Preparation. 1. From urine. Evaporate 1000 C.C. of
urine to near dryness, triturate the residue with clean
sand and add 60 C.C. of hydrochloric acid; finally,
extract with alcohol.   The acid alcoholic solution

is neutralized with soda ley, and evaporated to a
syrupy consistence with a small quantity of oxalic
acid, the residue dried in a water bath and treated
with a large quantity of ether containing 20 per
cent. of alcohol. When the residue is thoroughly
exhausted, the alcoholic etherial solution is evaporated and the crystalline residue treated with a
solution of milk of lime, and the resulting precipitate removed by filtration. The filtrate is concentrated and hydrochloric acid added, after standing
some hours hippuric acid will crystallize out. The
crystals are collected on a weighed filter, dried
and weighed; the weight gives the quantity of
hippuric acid in the amount of urine examined.

2. Artificially. By heating benzanide with chloracetic acid in a sealed tube at a temperature of
155°; thus

Chloracetic acid. Benzamide.   Hippuric acid.

$$C_2H_3ClO_2 + C_7H_5(H_2N)O = C_9H_9NO_3 + HCl$$

Chemical and physical characters. The crystals are
semi-transparent rhombic prisms; almost insoluble in cold water, soluble in boiling water and
alcohol, quite insoluble in ether, extremely soluble
in solutions of phosphate of soda. Heated at a
temperature of 250°, they are decomposed into
benzoic acid, benzamide, and hydrocyanic acid.
Boiled with strong hydrochloric acid, hippuric
acid is decomposed into glycocin and benzoic
acid.

Hippuric Acid.   Glycocin.   Benzoic Acid.

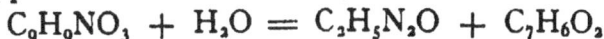

$$C_9H_9NO_3 + H_2O = C_2H_5N_2O + C_7H_6O_2$$

Hippuric acid is monobasic, the hippurates all give buff-coloured precipitates with ferric salts.

Tyrosin. $C_9H_{11}NO_3$.

Associated with leucin it has been obtained from all the glandular organs and secretions of the body.   So constant is the association of these two bodies, that they are considered to be products of the metamorphoses of the same kind of tissue. Dr. George Harley considers them to represent either the arrested or retrograde metamorphosis of glycocholic and taurocholic acids.

They are never found in normal urine, their presence in that secretion always denoting serious disease.

Preparation.   From Urine.   Precipitate the colouring and extractive matters with basic lead acetate, and filter; decompose the filtrate with sulphydric acid, and filter; the clear filtrate is to be concentrated and on cooling crystals of tyrosin wil be deposited.

Artificially.   By fusing albumin, casein, gelatin yellow elastic tissue, or epidermal appendages with an equal weight of potassium hydrate; after the reaction has continued some time, and the mass turns yellow, it must be dissolved in hot water.   This solution contains leucin and tyrosin, the latter is precipitated by the addition of acetic acid, and removed.   The impure crystals thus obtained are redissolved in boiling dilute hydrochloric acid, and the solution filtered through animal charcoal and precipitated by sodium acetate solution; the preci-cipitate is dissolved in a little ammonia and heated; on cooling, tyrosin will crystallize out.

Chemical and physical properties. The crystals are

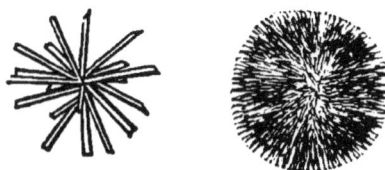

Fig. 3. Tyrosin.
a. Stellate crystal. | *b.* Spherical ball.

long prismatic needles which cluster together to form stellate groups; sometimes, when obtained from urine, these groups are so closely aggregated together as to form balls of spiculated needles. The crystals are sparingly soluble in cold water and alcohol; soluble in boiling water and in acid and alkaline solutions, the solubility being increased by the presence of extractive matters; insoluble in ether.

Tyrosin by distillation yields phenol. Warmed with a few drops of sulphuric acid, a solution of tyrosin gives after neutralization with barium carbonate, a violet reaction with ferric chloride. Treated with strong nitric acid, a yellow substance, nitrate of nitrotyrosin, is formed, which with hydrochloric acid gives a red, and with ammonia a brown colouration. A solution of tyrosin heated with a mixture of mercuric and mercurous nitrate, gives a red precipitate.

Tyrosin is sometimes regarded as amido propionic acid C_3H_5 $(NH_2)O_2$, in which 1 atom of H is
replaced by oxy-phenol C_6H_5O to make C_3H_4
(C_6H_5O) $(NH_2)O_2$.

LECITHIN. $C_{42}H_{84}PNO_9$.

This substance is a phosphoretted fatty body
originally obtained by Gobley from yolk of egg.
Mixed with cerebrin and oleophosphoric acid
it is obtained from the brain and other parts
of the animal organism. A mixture of lecithin
and cerebrin constitutes *protagon*, a viscous substance which dissolved in boiling alcohol is deposited in minute circular crystals when the liquid
cools; Liebreich considers protagon to be the chief
constituent of nervous tissue. The ultimate composition of lecithin, protagon, and cerebrin, is as
follows.

	Protagon.	Cerebrin.	Lecithin.
C.	67·4	68·45	64·27
H.	11·9	11·27	11·40
N.	2·9	4·61	1·80
P.	1·5		3·80
O.	16·3	15·67	18·73

A mixture of lecithin with cerebric acid and
cholesterin constitutes *myelin*; a fatty substance
met with in nervous tissue, blood corpuscles, and
crystalline lenses. When placed in water it
throws out peculiar spiral threads known as myelin
growths. Neubauer, however, does not consider
these myelin forms to be true growths but only

physical phenomena; as he found by adding a drop of ammonia to a drop of oleic acid, that on the addition of a drop of water these stringy threads at once appeared.

Hoppe Seyler believes that lecithin is often found in the body in combination with some albuminoid substance; thus, he considers vitellin to be a compound of globulin and lecithin.

Strecker believes that there are many varieties of lecithin, and many mixtures of these varieties.

The constitution of lecithin is very complicated, yielding on decomposition, glycero-phosphoric acid, oleic and palmitic acids, and cholin.

Preparation. Agitate the yolk of egg, or brain pulp, with a mixture of alcohol and ether, and afterwards filter. To the clear solution, add some alcoholic solution of platinic chloride containing a slight excess of hydrochloric acid; separate the yellow precipitate of lecithin platino-chloride by filtration. Decompose the precipitate with sulphydric acid, which liberates the lecithin hydrochlorate; treat this with moist silver oxide which removes the chlorine; the lecithin silver compound can now be decomposed with sulphydric acid, when silver sulphide will be formed and lecithin left as a waxy mass.

Chemical aud physical properties. Lecithin is an amorphous waxy substance; boiled with baryta water it is decomposed into glycero-phosphoric acid, oleic and palmitic acid, and cholin.

It is not decomposed by water, but is readily decomposed by weak alkalis and acids.

CEREBRIN. $C_{17}H_{33}NO_3$.

This substance mixed with lecithin, in different proportions, forms an important constituent of cerebral and nerve tissue; one of these compounds, protagon, we have already alluded to.

Preparation. Brain pulp is coagulated by heat; the coagulum treated with boiling alcohol, and the alcoholic solution filtered whilst hot. The precipitate which forms on cooling is separated and digested with a large quantity of ether for some days; the precipitate is then redissolved in boiling alcohol, and on cooling cerebrin is deposited in a pure state.

Chemical and physical properties. Cerebrin occurs as a white amorphous powder, without taste, insoluble in water and ether, freely soluble in boiling alcohol; neutral reaction. Heated with mineral acids it is decomposed; the products of this decomposition are not well understood, the chief appears to be a resinous oily substance. Pure cerebrin contains neither sulphur nor phosphorus in its composition.

Cerebric acid is probably another compound of cerebrin with lecithin, and is obtained by exhausting brain pulp with boiling alcohol, and allowing the mixture to stand for several days; the precipitated matters are then exhausted, first with cold and then with boiling ether. The etherial solution on cooling deposits cerebric acid and oleophosphoric acid; cold ether is added to remove the latter, and the cerebric acid dissolved in boiling ether will be deposited from the etherial solution on cooling in white crystalline grains, which are insoluble in cold alcohol and ether, and when treated with boiling swell up. Its elementary composition is

C. 66·7
H. 10·6
N. 2·3
P. 0·9
O. 19·5

GROUP II. PRIMARY DIAMIDES OR UREAS.

UREA. CH_2N_4O
KREATIN. $C_4H_9N_3O_2$
KREATININ. $C_4H_7N_3O$

UREA. CH_2N_4O.

This substance represents the last product of the oxidation in the body of albumin and the albuminous tissues, derived from the further oxidation of those imperfectly oxidized substances, xanthin, uric acid, kreatin, etc., which are the earliest products formed in the retrograde metamorphosis of albumin.

Is urea furnished equally by all the albuminous constituents of the body, or does it only represent the decomposition of one particular tissue or organ? This question has been differently answered by physiologists.

Formerly it was held that urea was derived

solely from the excess of the nitrogenous food taken into the system, but not employed, being at once oxidized and converted into urea: this view of its formation is known as the "luxus consumption theory." If urea were formed from the excess of food taken into the system, we should naturally expect to find urea disappear when food was withheld; this is not the case, since urea is always present in the urine of starving animals.

Other physiologists observed that kreatin was decomposed into urea in the blood, and as kreatin was abundantly found in the juice of flesh, they imagined that urea was derived solely from the disintegration and oxidation of muscular tissue, and that the quantity passed into the urine was a measure of the muscular activity. If this view were correct we should certainly expect to find a considerable increase of urea in the urine after muscular exercise; however, no such increase has been observed.

Professor Haughton found after a 5 mile walk, the urea excreted in the 24 hours was 501·28 grains, and after a 20 mile walk, the urea was 501·16 grains, or 0·12 grain less! Franque, after moderate exercise, excreted 588·7 grains, and after severe exercise 587· or 1·7 grain less! Similar observations have been made by Lehmann, Voit, Parkes, and Ed. Smith, proving that muscular activity has no special influence in the production of urea.

The view most generally adopted at present, regards urea as representing the metamorphosis of the whole of the nitrogenous elements of the

food which have undergone conversion into tissue.
Dr. Parkes has confirmed the observation of
Gilbert and Lawes that if the body be kept
perfectly quiescent, and varying quantities of
nitrogenous food given, the excretion of urea is
always in direct proportion to the nitrogen ingested. He has also shown, by a series of observations made on individuals, that all the nitrogen
taken as food was recoverable from the fæces and
urine, just the same, whether the body was at rest
or at work.

PARALLELISM BETWEEN THE ENTRANCE AND EXIT OF
NITROGEN. (Dr. Parkes.)

Case.	Age.	No. of days of experiment.	Nitrogen ingested.	Exit of Nitrogen. Urine.	Bowels.	Total exit.
1	28	26	270 grm.	254·04	27·74	281·78
		18	270 grm.	254·22	28·74	282·96
2	24	16	302·6	296·2	16·6	312·8

Ranke also confirms these observations, for he
found by experiment on himself that with an entry
of 296 grains of nitrogen, 26·23 grains were passed
by the bowels, and 281 grains by the kidneys,
making a total of 307·23 grains of nitrogen.

Dr. Parkes thus states his view of the formation
of urea.

"The albumin of the body is broadly divided
into fixed and circulating. All the albuminous
parts of muscles, nerves, gland-cells, membranes,
etc., belong to the former class; the serum of the
blood, and the fluids which come from it, to the
latter. Urea is derived principally, probably

solely, from the circulating albumin which has
once beeu fixed, but which has become unfit for
use, and has been detached from that organ of
which it formed a part. As to the method in
which the gland-cells form urea, we know as little
as we do of the mode of growth of cells, to which
the process is no doubt allied."   (Dr. Parkes
Croonian Lectures, 1871, published in Lancet.)

The average quantity of urea which passes into
the urine per diem may be stated at 30 to 40 grms.
The quantity, however, is subject to variation, depending much on the season of the year, the period of the day, the quantity and nature of the
food, the age, sex, body weight, and health of the
individual.

Dr. Ed. Smith found that more urea was passed
proportionately in summer than in winter, and
that the greatest hourly elimination was after the
morning and evening meals.

An animal diet greatly increases the elimination ;
thus Lehmann and Franque on a liberal animal
diet, excreted respectively 53·3 grms. and 91·2
grms. ; whilst on a mixed diet, the quantity was
reduced to 32·5 grms. and 38 grms. ; and on a
purely vegetable diet, a still smaller quantity was
excreted.

Taking large quantities of water, or using much
common salt with the food, increases the amount
of urea in the urine ; on the other hand tea, coffee,
starch, sugar, and alcohol, diminish the excretion.

Urea is excreted in larger quantities, in proportion to the body weight, in youth than in the adult,

whilst in old age it reaches its minimum. This is what we might expect, since in youth the chemical changes and tissue transformations are more active than in adult life. Men excrete little more urea in proportion to their body weight than women. Dr. Parkes gives the quantity of urea excreted in 24 hours as 3½ grains for every 1 lb. of body weight; consequently we may expect a larger elimination in heavy than in light persons. This relative proportion may hold in some cases, but, after a certain weight is attained, is not at all to be depended on.

In certain diseases the amount of urea excreted is greatly increased, in others it is diminished. In all febrile affections, and other acute diseases, which are accompanied with rapid disintegration of the tissues, the quantity is increased, and this increase is closely connected with the rise of temperature, indeed often precedes it.

The following experiment of Naumyn's shows this relationship between urea and the temperature. A dog was shut in an Obernier's chest, saturated with steam at a temperature of 35° to 40° for six hours; the dog's temperature rose to 42·5 and he excreted 110 C.C. of urine which contained 9·71 grms. of urea. Under ordinary conditions the dog's temperature was 39° and the quantity of urine passed in six hours was 120 C.C. which contained 7·3 grms. of urea. A reduction of temperature is followed by a reduction of the urea; thus, Schröder observed in a case of typhus that the urea daily eliminated averaged 41 grms.;

G

on the days, however, when the cold bath was employed, the urea fell to 33 grms. In another case the quantity decreased from 29·6 grms. to 19·9 grms.

In chronic diseases as phthisis (when hectic is absent), in cirrhosis of the liver, and especially in chronic albuminuria, the quantity is materially diminished.

When urea is retained in the system, a condition known as "uræmia" is induced. The exact nature of this condition has not yet been satisfactority determined. Some consider it to be due to the poisonous action of urea; others that the urea is decomposed into ammonium carbonate, and the blood poisoned with ammonia (ammonæmia); while others hold that excess of urea in the blood causes the water ¡of that fluid more readily to transude through the capillaries, and thus cause œdema of the brain. (Traube).

Preparation. 1. From urine. By evaporating the urine to a syrupy consistence, and treating it with nitric acid sp: gr. 1·25. The urea nitrate thus formed is decomposed by barium carbonate, and the mixture evaporated; the residue is treated with boiling alcohol, and filtered; the filtered solution yields, on cooling, crystals of urea.

2. Artificially. By heating ammonium cyanate

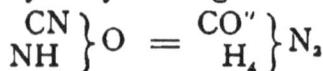

$$\left.\begin{array}{c}CN\\NH\end{array}\right\}O = \left.\begin{array}{c}CO''\\H_4\end{array}\right\}N_2$$

or by decomposing oxamide, with mercuric oxide;

Oxamide.   Urea.

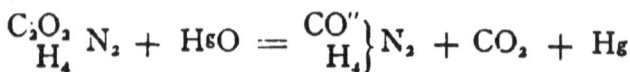

$$\left.\begin{array}{c}C_2O_2\\H_4\end{array}\right\}N_2 + H_6O = \left.\begin{array}{c}CO''\\H_4\end{array}\right\}N_2 + CO_2 + H_6$$

Chemical and physical properties. Urea crystallizes in colourless, four-sided prisms, which melt at 120°: they are soluble in their own weight of cold water, very soluble in hot water and alcohol. Its aqueous solutions are neutral to test paper. Heated to 150°, urea is converted into bi-uret and cyanuric acid. Heated in a sealed tube, with water, it is decomposed into ammonium-carbonate.

Urea unites with acids to form salts, of which the oxalate and nitrate are the most important.

Urea oxalate $(2CN_2H_4O,C_2H_2O_4)$: the crystals form long, transparent, tufted laminæ, composed of prisms and quadrilateral tables, they are very soluble in hot, but only slightly soluble in cold water.

Urea nitrate (CN_2H_4O,HNO_3): the crystals form shining, rhombic plates, soluble in 8 times their weight of cold water, more soluble in warm.

With mercuric oxide, urea forms an insoluble compound, which is formed in the quantitative estimation of urea with the mercuric nitrate solution. (For Quantitative estimation of urea, see urine.)

KREATIN. $C_4H_9N_3O_2$.

This substance is one of the primary products of muscle decomposition; and is always found in the juice of muscular tissue. In the blood it is decomposed either into urea or kreatinin. The proportion present in the flesh of different animals varies considerably; thus Liebig found in

* Fowls' flesh 3·2 per cent.
  Ox heart 1·37 ,,
  Pigeon ·82 ,,
  Beef ·69 ,, ′

The flesh of wild animals contains more kreatin than that of those kept in confinement. Large quantities of kreatin are obtained from meat extract, but as kreatin is an effete product, it has little nutritive value, the amount of force liberated in its conversion into urea and kreatinin being inconsiderable.

It is a question whether kreatin is a constituent of normal urine; whenever, however, it is not burnt off in the system into urea and kreatinin, it is found with the latter substance in the urine.

Preparation. From muscle. 200 grms. of finely chopped meat is mixed with an equal weight of water, and heated to 60°, the albumin being removed as it coagulates, the coagula pressed and the expressed fluid returned to the mother liquid. When all the albumin is completely coagulated and removed, the liquid must be set aside and filtered when cold. To the filtrate basic lead acetate is added, and the precipitate removed by filtration; the filtrate decomposed with sulphydric acid, and again filtered. This filtrate is evaporated to a syrupy consistence, and set aside for a few days, when crystals of kreatin will be deposited.

Chemical and physical properties. The crystals are prismatic, colourless and brilliant, having a bitter taste and giving a neutral reaction with test paper;

they are soluble in 80 parts of cold water; very
soluble in boiling water; slightly soluble in absolute alcohol, but insoluble in ether. Heated at 100°
they lose 12 per cent. of their weight from the
withdrawal of the water of crystallization.

Boiled with mercuric oxide, a powerful base
methyluramine is formed. Treated with either
sulphuric or hydrochloric acid kreatin is converted
into kreatinin, by the withdrawal of 1 molecule
of water; thus,

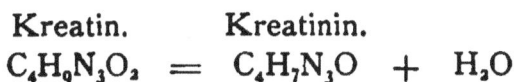

Kreatin. Kreatinin.

$$C_4H_9N_3O_2 = C_4H_7N_3O + H_2O$$

Boiled with baryta water, kreatin is decomposed
into urea and sarcosin; thus,

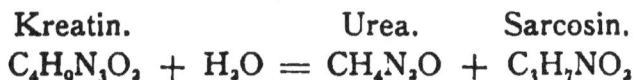

Kreatin. Urea. Sarcosin.

$$C_4H_9N_3O_2 + H_2O = CH_4N_2O + C_3H_7NO_2$$

KREATININ. $C_4H_7N_3O$.

This powerful base is constantly present in human urine; according to Neubauer the quantity
passed into the urine in 24 hours averages 0·6 to
1·3 grm. It is derived from the decomposition of
kreatin in the blood; in no case has it been obtained as a primary product of decomposition from
any of the tissues. Nawrocki has shown by recent experiments that it does not occur in muscular tissue either at rest or when tetanized.

Preparation. From urine. To 250 C.C. of the 24
hours urine, add milk of lime, till the urine has an

alkaline reaction; calcium chloride is then added till a precipitate ceases to be formed; filter and wash. Evaporate the filtrate and washings to a syrupy consistence and add 50 C.C. of alcohol, set the mixture aside in a cool place and filter off any precipitate that may form. The clear filtrate is then evaporated to 50 C.C., and when quite cold ½ C.C. of the alcoholic standard solution of zinc chloride (see appendix) is added; the mixture is set aside in a dark cool place, and after 24 hours, crystals of kreatinin zinc chloride will form. The crystals are collected on a weighed filter, washed with alcohol, till the washings are colourless, and the filter dried with the crystals between two watch glasses; and weighed. 100 parts of kreatinin zinc chloride represent 62·4 parts of kreatinin. To obtain kreatinin from the zinc compound; the latter must be boiled for some hours with an excess of hydrated lead oxide; the mixture is then filtered through animal charcoal, and evaporated to dryness, the residue treated with boiling alcohol, and the alcoholic solution concentrated; on cooling, the alcoholic solution will deposit crystals of kreatinin.

Chemical and physical properties. The crystals form oblique rhombic prisms, which are soluble in boiling water, soluble in 12 parts of cold water, and in 100 parts of absolute alcohol. It is an extremely powerful base, gives an alkaline reaction with test paper; and forms well defined basic double salts, as with zinc chloride, mercuric chloride and silver nitrate.

Boiled with potassium permanganate, kreatinin is decomposed into methyluramin oxalate.

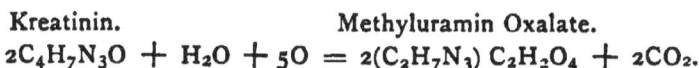

Kreatinin.   Methyluramin Oxalate.

$$2C_4H_7N_3O + H_2O + 5O = 2(C_2H_7N_3) C_2H_2O_4 + 2CO_2.$$

GROUP III. SECONDARY DIAMIDES OR UREIDES.

$$Diureides* \begin{cases} \text{Uric Acid} & C_5H_4N_4O_3 \\ \text{Guanin} & C_5H_5N_5O \\ \text{Xanthin} & C_5H_5N_4O_2 \\ \text{Hypoxanthin} & C_5H_4N_4O \\ \text{Allantoin} & C_4H_6N_4O_3 \end{cases}$$

Uric Acid. $C_5H_4N_4O_3$.

Uric acid is always present in small quantities in human urine, it has also been found in the blood, the spleen, and liver. In birds, reptiles and insects whose tissue metamorphosis, is represented by this product rather than urea, the semi-solid excrement is almost entirely composed of ammonium urate. .

Uric acid represents one of the products of the " retrograde metamorphosis" of the albuminous constituents of the body: and in all mammalia is itself decomposed in the blood into urea and carbonic acid; urates introduced into the system increasing the urea in the urine.

The quantity of uric acid daily eliminated with

* Contain two residues of urea and one residue of a non-nitrogenous acid.

the urine varies with the nature of the food, the amount of exercise taken, and the season of the year. Thus Lehmann has shewn, that with a purely animal diet the daily excretion was 1·47 grms; with a mixed diet, the excretion was 1·18 grms: and with a vegetable diet 1·02 grms. Wine, beer, and spirits likewise increase the excretion. Hammond in experiments on himself found that with moderate exercise he passed daily, on an average, 0·86 grammes or about 13 grains; and with increased exercise 0·55 grammes or about 8 grains. In animals confined in cages, Lehmann found the uric acid increased considerably the same has been noticed with human prisoners.

Dr. G. Harley states that the excretion of uric acid is greatly increased in winter.

Dr. Parkes gives 0·555 grammes or 8½ grains as the new average of uric acid passed by the human adult in the twenty-four hours. And Ranke gives the relative proportions of uric acid to urea, as 1 : 50 and 1 : 80.

Uric acid is never found free in normal urine, but always in combination with soda, potash and ammonia; forming soluble salts, the *urates*. Whenever free uric acid, or a deposit of urates occur in the urine, some abnormal condition of the system or of the urine is indicated.

These deposits may be met with under the following conditions. 1st. An excess of uric acid may be passed into the urine. 2nd. The quantity of water of the urine may be diminished, there being no excess of uric acid, and; 3rd. They may be thrown down by an extremely acid condition of the u. ne.

1. Uric acid, as before stated, is converted before leaving the system, by a process of final oxidation, into urea and carbonic acid. Whenever this process of oxidation is incompletely performed, uric acid passes out of the body in greater quantity; and we may consequently expect to find an excess of it in those diseases which impede the respiratory function, as phthisis, pneumonia, emphysema and chronic affections of the heart; and i.: those diseases which impair the oxidizing power of the blood, as leucocythæmia and chronic diseases of the liver and spleen.

2. In some conditions of the system, as febrile excitement, and unusual amount of exercise attended with profuse perspiration, the quantity of water eliminated by the kidneys is much diminished, and consequently the urine is nearly saturated with the uric acid salts, which are then only retained in solution as long as the urine remains warm, and as soon as the urine cools they are deposited.

3. The presence of free acid in the urine decomposes the soluble neutral salts of uric acid, forming acid urates which are less soluble, or liberating the insoluble uric acid. As the acid phosphate of sodium does not effect this decomposition, it is probable that the free acid is furnished by an acid fermentation of the pigmentary matters of urine.

The urine during an attack of gout contains a smaller proportion than usual of uric acid; in this case, however, the uric acid is retained in the system and saturates, so to speak, the serum of the blood, and the vascular and non-vascular tissues of the body. That this is the case was shown by Dr. Garrod who obtained crystals of uric acid deposited on pieces of fine thread from the serum, (acidulated with acetic acid), of a gouty patient.

In cases of chronic gout, concretions, popularly called chalk-stones, form in the joints, and on the lobes of the ears, these concretions consist of sodium urate.

Preparation. Uric acid is obtained abundantly from the white semi-solid excrement of birds and reptiles. Dissolve 1 drachm of serpents' dung in 8 ounces of water, and boil with an excess of potash; when ammoniacal fumes are no longer given off, filter the solution; to the brown solution thus obtained add hydrochloric acid in excess, when, on standing, white crystals of uric acid will be deposited; purify by washing with cold water, redissolving in potash, and reprecipitating with hydrochloric acid, even the crystals thus obtained are not yet free from colouring matters.

Quantitative estimation in urine. (See Urine.)

Chemical and physical properties. The crystals of uric acid obtained from human urine present a great variety of forms; the most common are the smooth, transparent, rhomboidal·tables of variable size; mixed with these are diamond-shaped plates, hexagonal tables, rectangular, four-sided prisms, bundles of needle-shaped crystals, saw-like and dentated. If these mixed crystals are dissolved in liquor potassæ, and recrystallized by the addition of hydrochloric acid, the characteristic rhomboidal tables become more evident.

Fig. 4. Uric Acid.

The crystals are extremely insoluble, requiring 15,000 pts. of cold water to dissolve them; they are quite insoluble in ether and alcohol.

Moistened with nitric acid, and gently evaporated, a reddish residue is obtained, which treated with ammonia, gives a magnificent violet-red. (murexide test).

A drop of uric acid solution, with sodium carbonate, allowed to touch a paper prepared with silver solution, gives a brown stain.

One part of uric acid, treated with four parts of nitric acid (sp. gr. 1·42) gives a substance which stains the skin pink, called *alloxan*.

Uric acid is at once thrown down from its solution by the addition of a free acid, mineral or organic.

Uric acid is dibasic; and forms with bases both acid and neutral salts, which are sparingly soluble in cold water, more soluble in hot, and still more soluble in alkalis. The acid are less soluble than the neutral salts. Potassium, Sodium and Ammonium urates are always present in human urine; the calcium salt is, however, rarely met with.

GUANIN. $C_5H_5N_5O$.

This substance is a normal constituent of the semi-solid excrement of birds; and is obtained in large quantities from guano, which is the dried excrement of sea-fowl, found in vast beds in the islands of the Pacific Ocean. It is also obtained from the excrement of the garden spider (epeira diadema). It has sometimes been met with in the human liver, spleen, and fæces, but but does not appear to be a constant product.

Preparation. Guanin may be extracted from the substance which contains it, by boiling with milk of lime, filtering the hot solution, and neutralizing with hydrochloric acid; on standing for some time, guanin mixed with uric acid will be deposited. This deposit is to be treated with boiling hydrochloric acid, which dissolves the guanin but not the uric acid; the hot acid solution of guanin is filtered, and on cooling crystals of guanin hydrochlorate separate; these are decomposed by ammonia, and the base guanin set free.

Chemical and physical properties. Guanin is a yellowish-white, amorphous substance, nearly insoluble in water, ether, and alcohol, but soluble in dilute acids and alkalis. Treated with hydrochloric acid and potassium, it is decomposed into guanidin and parabanic acid.

Guanin.   Parabanic.   Guanidin.

$$C_5N_5H_5O + H_2O + 3O = C_3N_2H_2O_3 + CN_3H_5 + CO_2$$

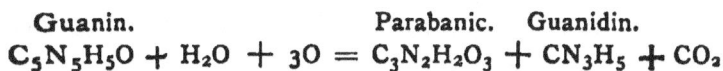

Acted on by nitrous acid, guanin is transformed into xanthin with the liberation of nitrogen; thus,

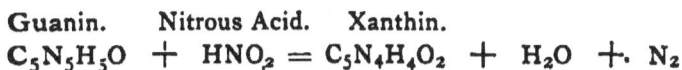

Guanin.   Nitrous Acid.   Xanthin.

$$C_5N_5H_5O + HNO_2 = C_5N_4H_4O_2 + H_2O + N_2$$

By oxidation with potassium permanganate, guanin is converted into urea, oxalic acid, and *oxy-guanin*.

Xanthin. $C_5H_4N_4O_2$.

This substance can be obtained from many of the tissues and secretions of the human body, as the liver, spleen, thymus gland, muscle, and the blood. It is not however found in normal urine, as in a healthy condition it undergoes immediate oxidation in the system and is converted into other products. Xanthin is a constituent of certain rare urinary calculi, and Dr. Bence Jones has recorded an interesting case of xanthin gravel occurring in a lad aged nine and a half years. The xanthin calculus removed by Langenbeck was also from a boy. Dr. Bence Jones considers that the xanthin diathesis will be found generally to occur in youth, as it is in the early period of life the greatest chemical variations of the body are to be expected, and the imperfect oxidation of xanthin into uric acid most likely to occur.

Preparation. 1. From guanin. By oxidation with nitrous acid.

2. From urine. Add baryta water to the urine

supposed to contain xanthin, till a precipitate is no
longer thrown down; filter, and evaporate the
filtrate to a syrup, and allow it to crystallize. The
mother liquor, after the removal of the crystals,
is boiled with cupric acetate, and the precipitate
thus formed is removed by filtration, washed, and
dissolved in warm nitric acid. This acid solution
is precipitated by silver nitrate, and the resulting
precipitate washed and crystallized from hot dilute
nitric acid, and the crystals washed with ammoniacal silver solution, and suspended in water.
The aqueous solution is to be decomposed with
sulphydric acid, filtered, and the filtrate evaporated;
the residue yields xanthin.

Chemical Properties. Xanthin forms white scales,

Fig. 5. Xanthin.

somewhat resembling bees-wax in appearance.
Deposited spontaneonsly from urine it occurs in
lemon shaped plates, (a) these dissolved in hydrochloric acid and the solution evaporated, yield
prismatic and hexagonal crystals. (b.) Xanthin is
insoluble in water, alcohol, and ether; soluble in
alkaline solutions from which it is deposited by a
current of carbonic acid gas, and in strong mineral acids. By the action of sodium amalgam
xanthin is converted into hypoxanthin.

Xanthin.   Hypoxanthin.

$$C_5H_4N_4O_2 \; - \; O \; = \; C_5H_4N_4O$$

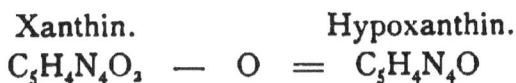

Burnt in air, it gives off the odour of scorched hair.

Evaporated with nitric acid on platinum foil and the residue moistened with liquor potassæ, it yields a dark purple colour.

Xanthin gives white precipitates, with mercuric chloride and silver nitrate. Dissolved in hydrochloric acid, it gives with platinic chloride a yellow crystalline precipitate.

HYPOXANTHIN. $C_5H_4N_4O$; (syn. Sarcine).

Hypoxanthin is found in the juice of flesh, in the spleen, and in the thymus and thyroid glands. As a morbid product, the result of deficient oxidation, it has been obtained from the blood and urine of leukæmic patients. .

Preparation. An aqueous extract of muscular tissue, or spleen pulp, is precipitated with baryta water, and the precipitate filtered off. To the filtrate ammoniacal solution of silver nitrate is added, and the greyish white precipitate collected on a filter, and washed. The precipitate is then to be suspended in water, decomposed with sulphydric acid, the mixture boiled for some time, and filtered while hot. The filtrate is next evaporated to dryness, and the residue contains uric acid, xanthin, and hypoxanthin. To separate the uric acid and xanthin, the residue is to be dissolved in dilute sulphuric acid, boiled and filtered while hot; to the filtrate, when cold, mercuric nitrate is added and filtered; to the filtrate some ammoniacal solution of silver nitrate is added, the precipitate consists of hypoxanthin nitrate and silver oxide; this is to be decomposed with sulphydric acid, and hypoxanthin is precipitated.

Chemical properties. Hypoxanthin is a white, imperfectly crystalline powder, rather more soluble in water, and alcohol, than xanthin; it dissolves freely in acid and alkaline solutions.

Its nitric acid solution gives a copious white precipitate with nitrate of silver.

At high temperatures it is decomposed, forming hydrocyanic acid, and cyanuric acid.

ALLANTOIN. $C_4H_6N_4O_3$.

This substance is obtained from the allantoic fluid of the cow, and from the urine of young sucking animals. It has been found in the urine of animals whose respiration has been impeded; and Schottin has recorded a case in which he met with it in the urine of a man who had taken a large quantity of tannic acid.

Preparation. 1. Artificially. By boiling uric acid with lead peroxide; thus,

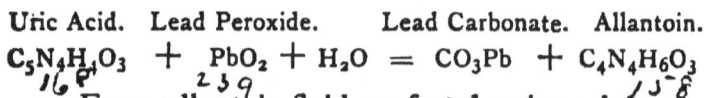

Uric Acid.　Lead Peroxide.　　Lead Carbonate.　Allantoin.

$$C_5N_4H_4O_3 + PbO_2 + H_2O = CO_3Pb + C_4N_4H_6O_3$$

2. From allantoic fluid or fœtal urine; by evaporating the fluids to one-fourth their volume, and allowing crystals to form by standing in the concentrated solution. The crystals are purified

by redissolving with a little warm water slightly acidulated with hydrochloric acid, and filtering the hot solution through animal charcoal;⁰ on cooling pure crystals of allantoin will be deposited.

Chemical and physical properties. The crystals are colourless, hard, glassy prisms, of neutral reaction; soluble in 160 parts of cold water, very soluble in boiling water. Heated in air they are decomposed, furnishing ammonium cyanide and carbonate. Boiled with potash it yields potassium oxalate.

* For such purposes animal charcoal must be freed from mineral matters by digestion in dilute hydrochloric acid and thorough washing.

PART III.

INORGANIC CONSTITUENTS OF THE BODY.

CHAPTER VI.

THE inorganic substances occurring in the animal tissues and fluids are not chemically combined with the organic principles,* but are as it were held in solution, and can readily be removed by mechanical means. For example, if some tissue as muscle be minced very fine and placed on a dialyser, and the dialyser floated on water, after the lapse of some hours, the flesh will lose its consistence and become pulpy and jelly-like, at the same time the water will increase in density from the diffusion of the inorganic salts into it; and these salts can be obtained for analysis by evaporating the diffusate.

The inorganic substances enter and pass out of the system as crystalloids, but whilst in contact with organic matter they seem to lose their crystalline form and become colloidal; and thus give the tissues and fluids their homogenous appearance.

The inorganic matters are introduced into the

* The only exception to this statement is, the chemical combination of albumin with potassium and sodium to form albuminates.

system with the food and drink, and the rapidity with which they at once pass into the textures is astonishing. Dr. Bence Jones has shown, by experiment, that in the human body twenty grains of lithium carbonate taken into the stomach will in two and a half hours have passed into the textures; and in three and a half hours it will be distinctly present in every particle of the crystalline lens, thus passing beyond the blood circulation into the most distant parts. The same authority states that the lithium carbonate is retained in the system for six, seven, or eight days before it is entirely eliminated.

The various salts (with the exception of sodium chloride) pass through the system unaltered and are recoverable from the urine and fæces. In the case of sodium chloride, only four fifths of that taken into the system passes out as such; the remaining fifth being decomposed by acid potassium phosphate to form potassium chloride and acid sodium phosphate; thus,

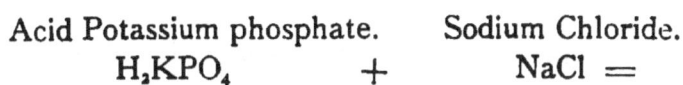

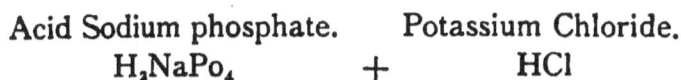

Acid Potassium phosphate.   Sodium Chloride.

$$H_2KPO_4 \quad + \quad NaCl =$$

Acid Sodium phosphate.   Potassium Chloride.

$$H_2NaPo_4 \quad + \quad HCl$$

The inorganic principles subserve two important offices in the economy; viz.

1. Mechanical, in giving strength and firmness to those textures which, like bone, cartilage and muscle form the solid portion of the organism.

2. Chemical, in effecting certain metamorpho-

ses in the tissues and fluids ; and keeping in solution many of the otherwise insoluble organic
principles.

*The determination of the amount of water present in a
tissue or fluid.* A definite quantity of the tissue
(finely divided) or fluid is placed in an accurately
weighed platinum capsule and evaporated to dryness on a water bath, till it no longer loses weight.
The capsule and the dry residue are then weighed,
and the loss of weight will represent the quantity
of water present in a certain weight of the substance examined.   For example, if 60 grm. of
blood are evaporated and the dry residue weighs
13 grms. then the blood has lost 47 grms., or the
quantity of water present in 60 grms. of blood ; to
find from this the quantity of water present in 100
parts it is only necessary to make a simple proportion ; thus,

Weight of blood.   Loss of weight.   Proportion of water in 100 parts.

$$60 \quad : \quad 47 \quad :: \quad 100 \quad : \quad 78\cdot3$$

*Estimation of the organic and inorganic constituents of
the tissues and fluids.* The dry residue, left after the
evaporation of the water, represents the organic
and inorganic materials of the blood.   To estimate
these, the platinum capsule now fitted with a cover is
burnt over a Bunsen's lamp till the ash is of a greyish red colour ; the platinum capsule is then with- ,
drawn and when cold weighed.   The weight lost
by incineration represents the quantity of organic
matter in 60 grms. of blood ; and the weight of
the residue left after incineration represents the

amount of inorganic matters; thus, if the residue left after the evaporation of 60 grms. of blood weighs 13 grms. and after incineration weighs only 5 grms, it is evident that 8 grms. of organic matter have been burnt off, and that 5 grms. and 8 grms. respectively represent the quantity of inorganic and organic principles present in 60 grms. of blood: and to find the proportion of these substances in 100 parts it is necessary to calculate as follows:—

Weight of blood.	Weight after incineration.	Weight of inorganic matter in 100 parts.
60	: 5 ::	100 : 8·3

and

Weight of blood.	Weight lost by incineration.	Weight of organic matters in 100 parts.
60	: 8 ::	100 : 13·3

THE ALKALINE CARBONATES.

THE SODIUM AND POTASSIUM CARBONATES are abundantly diffused throughout the textures; and with the alkaline phosphates maintain the alkalinity of the blood. Their principal use in the system is to retain in solution the loose carbonic acid derived from the disintegration of the tissues till it is exhaled from the lungs. In the capillaries of the lungs the carbonates are themselves decomposed, and the combined or stable carbonic acid is set free, and is exhaled with the loose carbonic acid; how this decomposition is effected is not exactly known, but Dr. Preyer has suggested that it is effected in some way by the union of oxygen with hæmoglobin.

The alkaline carbonates are principally derived from the oxidation of the tartrates, citrates, malates, etc., introduced into the system with our vegetable food; hence the carbonates are more abundant in the blood of herbivora than of carnivora: also after the administration of any of the above acids, the urine speedily acquires an alkaline reaction. The carbonates are also formed by the oxidation of those acids, as acetic, lactic, etc., which are formed within the system by the decomposition of the tissues, and which are speedily reduced to CO_2, when they pass into the general current of the circulation.·

CALCIUM CARBONATE. This salt associated with calcium phosphate is found in the bones, in the relative proportion of 1 part of calcium carbonate to 5 parts of calcium phosphate: it is also found in a crystalline state in the saccule of the vestibule of the internal ear.

In certain diseases of bone, as rachitis, osteo malacia, and caries, calcium carbonate is met with in the blood and urine, being retained in solution, when in small quantities, by the alkaline chlorides and free carbonic acid. When however its elimination through the kidneys is checked, owing to disease of those organs, or the quantity removed from the bones to the blood excessive, it is deposited in the tissues, and the stomach, lungs, intestines, and kidneys become coated with this salt.

Local deposits of calcium carbonate are frequently met with in tissues which have suffered from a general impairment of vitality or from diminished nutritive supply. In these cases, it is probable, that owing to the escape of the free carbonic acid, which keeps this salt in solution, from the stagnated fluids of the part, the calcium salt is precipitated, and on account of the degeneration of the tissues is not taken up by them.

Determination of Carbonic acid. A weighed portion of the carbonate is dissolved in water and introduced into the flask (*a*); this flask is fitted with a bulb tube (*c*), filled with dilute nitric acid, which is prevented from flowing into the mixture by means of a pinchcock (*d*).

The apparatus is now weighed, and attached to (*b*), a flask containing concentrated sulphuric acid, and the bulb tube (*c*) pushed down to nearly the bottom of flask (*a*), the pinch-cock pressed and the dilute nitric acid allowed to mix with the contents of the flask. When the carbonate is completely decomposed, the flask (*a*) is placed in warm water and gentle suction applied to tube (*f*) till all the carbonic acid is removed; the apparatus is then allowed to cool and is again weighed: the loss of weight represents the amount of carbonic acid present in the salt examined.

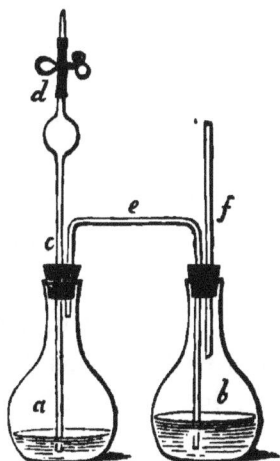

THE PHOSPHATES.

THE EARTHY PHOSPHATES. The calcium phosphate (Ca_32PO_4) is the principal inorganic constituent of bone, enamel, dentin, and cartilage; it is also found in smaller quantities in every other texture and fluid of the body.

QUANTITY OF CALCIUM PHOSPHATE IN 1000 PARTS.

Enamel of teeth	. 885	Muscle	. 2·5
Dentin . . .	. 643	Blood	. 0·3
Bone	. 550	Gastric juice . .	. 0·4
Cartilage . .	. 40		

Magnesium phosphate associated with calcium phosphate is found in all the animal tissues and fluids, but in smaller quantities; the proportion in 1000 parts of bone being about 13. In the excreta however the magnesium is relatively more abundant than the calcium salt; from this it may be assumed that the magnesium salt is less required by the organism, and is consequently not so long retained by it as the calcium salt, and also that less is absorbed by the intestinal canal.

In the body, the calcium phosphate is retained in solution by the dissolved albuminous matters; the magnesium phosphate by the alkaline chlorides and phosphates.

In the urine both salts are held in solution by the acid sodium phosphate (H_2NaPO_4) and so long as the urine remains acid no precipitate occurs; but when that fluid becomes alkaline, either by the

addition of a few drops of ammonia, or by the decomposition of the urea, they are at once deposited.

The average quantity of earthy phosphates passing out of the system into the urine in the 24 hours,
may be placed at from ·9440 grms. to 1·012 grms.;
of which the magnesium salt constitutes about 67
per cent, and the calcium salt 33 per cent.

THE ALKALINE PHOSPHATES. The potassium and
sodium phosphates are distributed extensively
throughout the body, being readily dissolved in
the tissues and fluids owing to their great solubility.

The potassium and sodium phosphates form
three varieties of salts; viz., an acid sodium or
potassium phosphate, as H_2MPO_4; a neutral phosphate, as HM_2PO_4; and an alkaline sodium or
potassium phosphate, as M_3PO_4. It has already
been stated that some portion of the sodium chloride is decomposed by the acid potassium phosphate into potassium chloride and acid sodium
phosphate, the latter salt is abundantly present in
the urine, and is one of the chief causes of the acid
reaction of that fluid. The alkaline sodium and
potassium phosphates with the alkaline carbonates
give to blood its alkaline reaction; the alkaline
condition of the blood promotes the oxidation of
the albuminous constituents, and is necessary for
the due performance of its vital functions.

The quantity of alkali-phosphates passed into
the urine in the 24 hours may be stated, as varying
from 1·40 grms. to 2·30 grms.

· *Determination of Phosphoric acid.* Dissolve the
ash in a small quantity of dilute acetic acid, and

proceed as directed for the quantitative estimation of phosphoric acid in urine.

THE SULPHATES.

The sulphates are derived from the food, and from the oxidation of the sulphur of the albuminous constituents within the body. Very little is known regarding the part they play in the economy. As their elimination is increased by exercise, febrile excitement, etc., they must be regarded as products of increased tissue metamorphosis. About 2· grms. are passed into the urine in the 24 hours.

Determination of Sulphuric acid. Dissolve the ash in water, and proceed as directed for the quantitative estimation of sulphuric acid in urine. (See Urine.)

THE CHLORIDES.

The sodium and potassium chlorides are universally diffused through the different tissues and fluids of the body ; the sodium salt, however, being more abundantly met with in the blood, and the potassium salt in muscular tissue.

The chlorides have the following uses in the economy. In the first place, they aid in keeping the albuminoid principles in solution. Secondly, they regulate the endosmotic action in different parts of the body, thus Liebig considers the presence of common salt in the blood to be one of the main causes of the phenomenon of absorption. Thirdly, they exercise an important influence on the development and growth of the body ; the instinctive craving for this article of diet shows how important it is to the economy, for animals deprived of the use of salt speedily fall into bad condition, and will travel miles to obtain it; and certain African tribes, in districts where salt is scarce, will even barter gold for an equal weight of this commodity.

Sodium chloride is found in great abundance in all cellular growths, as in cartilage, mucus, etc. In certain diseases attended with increased cell formation, as in cancer, in pneumonia with exudation, and purulent discharges, sodium chloride is present in large quantities in the morbid products, and consequently there is a sensible diminution in the quantity excreted by the urine.

The average quantity of chlorides passed into the urine in the 24 hours may be put at 5 to 8 grms. ; varying of course with the quantity ingested.

Determination of Chlorine. Dissolve the ash in water, and proceed as directed for the Quantitative estimation of chlorides in Urine.

POTASSIUM AND SODIUM.

The salts of potassium and sodium are found to have different relative proportions in the different tissues and fluids ; thus, the potassium salts are relatively more abundant than the sodium salts in muscular tissue, and the sodium salts in blood.

PROPORTION OF POTASSIUM AND SODIUM SALTS IN 100
PARTS OF THE ASH OF MUSCLE AND BLOOD.

	Potassium.	Sodium.
Muscle .	58	23
Blood .	6	79

Determination of Potassium and Sodium. The ash
is thoroughly exhausted with boiling water, and a
solution of ammonium carbonate added till no
further precipitate is thrown down ; the mixture
is then filtered, and the precipitate well washed.
The filtrate and washings are acidulated with
hydrochloric acid, and evaporated, in a platinum
capsule furnished with a lid, to dryness. The
residue is dissolved in a little water with a few
drops of ammonia and ammonium carbonate, filtered, and the filtrate again evaporated. The
Potassium and Sodium Chlorides thus obtained,
are weighed, dissolved in a little water, an excess of platinum bichloride added, and the mixture evaporated to dryness. The residue is exhausted with spirits of wine, allowed to stand
some hours, and the potassio platinum chloride
removed by filtration, dried, washed with alcohol,
again dried, and then weighed.

100 parts of the potassio platinum chloride represent 30·51 parts of potassium chloride, or 16
parts of potassium.

The weight of the potassium chloride deducted
from the whole weight of the alkaline chlorides

gives the quantity of sodium chloride: 100 parts of sodium chloride represent 39·34 of sodium.

LIME AND MAGNESIA.

The carbonates and phosphates of lime and magnesia are the chief constituents of bone; 100 parts of bone yielding about 64 per cent. of these salts. They are present, but in much smaller quantities, in most of the other tissues.

Determination of Lime and Magnesia. The ash is dissolved in a little water acidulated with acetic acid, and filtered: to the filtrate ammonium oxalate is added as long as a precipitate is formed, and the mixture set aside in a warm place till the whole precipitate is deposited. The precipitate (calcium oxalate) is now transferred to a platinum crucible and ignited; when cool, the residue is moistened with a few drops of sulphuric acid and the caustic lime converted into calcium sulphate.

100 parts of calcium sulphate represent 41·18 parts of CaO; and 3 equivalents of calcium sulphate, $CaSO_4$, are equal to 1 equivalent of calcium phosphate Ca_32PO_4.

The filtrate from which the calcium oxalate has been removed, is now treated with an excess of ammonia, which precipitates the ammonio magnesium phosphate. This precipitate is allowed to settle, and is then collected on a filter and washed with a dilute solution of ammonia; the precipitate is now dried, and the filter incinerated on the lid of the crucible. When the combustion of the filter is complete, the ash is placed in the crucible, the lid inverted, and submitted to an intense red heat; which converts it into a white shining mass of magnesium pyrophosphate $Mg_2P_2O_7$.

100 parts of magnesium pyrophosphate represent 35·94 parts of magnesia MgO.

IRON.

Traces of iron are met with in the ash of most tissues and fluids; as a proto chloride in the ash of gastric juice, and as a phosphate from muscular and splenic juice. Combined with some albuminous substance, it is present in all animal pigments.

Determination of Iron. The ash must be dissolved in hydrochloric acid, and heated; to the acid solution a little sodium sulphite is added and the mixture boiled till the liquid is colourless, diluted with distilled water and cooled. A Mohr's burette then filled with a standard solution of Potassium Permanganate,° which is added in small quantities to the colourless solution of iron, agitating after every addition the vessel that contains the mixture. When the operation is completed the iron solution acquires a pale rose red which does not disappear on agitation. The number of C.C.'s of the standard solution used from the burette must now be taken; and as 1 C.C. of the Permanganate solution corresponds to, ·0005 grm. of iron, then if 5 C.C. of the·standard solution be used, the quantity of iron contained in the ash will be ·0025 grm.

* See Appendix.

Silicon and Fluorine.

These substances are present in extremely minute quantities in the human body.

Silica is a constituent of the epidermal tissues, it is nearly always present in the fæces, and occasionally in the blood, bile and urine. Silica may be obtained from the ash of any of the substances in which it is present; by fusing the ash with 8 times its weight of sodium carbonate, and boiling the mass in water, on the addition of hydrochloric acid the silica is partly precipitated as a gelatinous mass; the acid solution is now evaporated, and the residue treated with some more hydrochloric acid and dried; the silica will then be left as a white insoluble powder.

Fluorine, united with calcium, as calcium fluoride, is found in minute quantities in the bones and teeth, also in blood, milk, and urine. The presence of fluorine in any of the textures can be readily demonstrated by boiling some of the ash with a little dilute sulphuric acid in a test tube, when the inner surface of the glass will become eroded from the liberation of hydrofluoric acid.

Copper, Lead, and Arsenic.

These substances are only incidentally present in the tissues, and are in no way necessary to the maintenance of their functions.

PART IV.
TISSUES AND FLUIDS.

CHAPTER VII.
THE SOLID TISSUES OF THE BODY.

CONNECTIVE TISSUE.   (Syn. fibrous, areolar.)

THIS tissue, as its name implies, connects the various organs of the body together; it thus forms the basis of the ligaments, the tendons, the fascia, the aponeuroses, the skin, and the adipose, serous and mucous tissues.   Under the microscope it appears as bundles of wavy filaments mostly running in a parallel direction.   By long boiling it yields gelatin.   Acetic acid causes it to swell up and become transparent, and lose its fibrillar appearance, at the same time bringing into view certain granular bodies, and a few filaments of elastic tissue.   Acetic acid does not dissolve it unless heated with it, and the solution thus formed is not precipitated by potassium ferrocyanide.

ELASTIC TISSUE.

Connective tissue always contains a few filaments of elastic tissue, and in some cases they predominate, as in the yellow elastic ligaments of the vertebræ, the sarcolemma of the muscular fibres, and neurilemma of the nerves.   It does not yield gelatin by boiling, does not swell up on the addition of acetic acid. By successively boiling with alcohol, ether, water, concentrated acetic acid, dilute alkalis, hydrochloric acid, it yields Elasticin.

EPIDERMAL TISSUE.

Horn, hair, wool, nails, hoofs, scales, feathers, and epithelium of the skin and mucous membrane are formed of this tissue; it is composed of a number of layers of cells and nuclei, which are rendered visible by treatment with strong solutions of caustic potash. It yields a substance called keratin.

ADIPOSE TISSUE.

Is formed by the accumulation of fat cells in the meshes of connective tissue. These fat cells are spheroidal sacs composed of a delicate membrane containing fatty matter, which during life is fluid; after death the contents become solid by the separation of the stearin and palmitin from their solution in the fluid olein, sometimes crystals of these substances are found in the interior of the cells. If agitated with ether, the fat will be dissolved out, leaving the cell membrane behind. Stearin and palmitin constitute three-fourths, and olein one-fourth of human fat. Adipose tissue is found in layers beneath the skin, or aggregated in masses in the cavities of the body surrounding certain organs, and everywhere filling up spaces between layers of muscle, vascular canals, etc. It is absent

I

in the brain, liver, and lungs, also in the fine skin of the eyelids, scrotum, and prepuce.

CARTILAGINOUS TISSUE.

Is met with in two states, the temporary and permanent. The former exists chiefly in the fœtus, and forms the foundation of the future bones. The latter covers the joints and is also met with in certain special cartilages as those of the ear, the nose, the thyroid and the epiglottis, it forms part of the fibro cartilages of the symphysis pubis and intervertebral substance.

It consists of a uniform matrix or intercellular substance, containing more or less fibrous tissue, in which certain nucleated cells are imbedded.

By long boiling the matrix yields chondrin, which is precipitated by acetic acid insoluble in excess, and by alum which redissolves if added in excess. The cells do not yield chondrin, but a substance analogous to it, which is precipitated by alum but not redissolved in excess.

Cartilage, taken from the ribs, contained 65 per cent of water, 34.40 of organic and 4·40 of inorganic matter; the latter comprising 4·0 parts of calcium phosphate, ·12 of magnesium phosphate, and ·28 of sodium chloride.

OSSEOUS TISSUE.

Is formed of a gelatinous matrix impregnated with calcareous salts, the quantity of which varies in different bones and at different ages. In certain diseased conditions, as rachitis, mollities ossium, and caries, the calcareous constituents are diminished.

The following table represents the average composition of healthy human adult bone in 100 parts.

Water and Organic Matter 33·30
Calcium phosphate   51·04
Calcium fluoride   2·00
Calcium carbonate   11·30
Magnesium phosphate   1·16
Sodium Chloride   1·20

WATER. Bone contains less water in proportion to its solids than any other tissue, except enamel and dentine, pulverized bone evaporated to dryness losing only 10 per cent of its weight.

ORGANIC MATTER. A fragment of bone digested for 36 hours in dilute hydrochloric acid has all the inorganic matters removed, leaving the gelatinous matter or ossein intact, this after boiling some hours in water dissolves, and is converted into gelatin.

INORGANIC MATTER. Incinerate some crushed bone in the muffle furnace to burn off the animal matter, and divide the residue into three portions

(a) Dissolve in a little dilute hydrochloric acid and neutralize with ammonia; a precipitate of *earthy phosphate* is thrown down.

(b) Boil with a little dilute sulphuric acid in a test tube; the inner surface of the glass will become eroded from the liberation of hydrofluoric acid, from the *calcium fluoride*.

(c) Boil with some water and filter, to the filtrate add a few drops of silver nitrate solution ; a white precipitate, indicating the presence of *chlorides* will be thrown down.

Fresh powdered bone treated with dilute hydrochloric acid effervesces, and with the acid solution ammonium oxalate gives a precipitate of calcium oxalate; this test indicates the presence of *carbonic acid and lime.*

DENTAL TISSUE.

The teeth are composed of three substances, termed respectively dentine, enamel, and cement.

Dentine forms the chief bulk of the tooth ; it is a tubular structure, and differs from bone only by containing less animal matter and in having no lacunæ or proper canaliculi. The proportion of organic matter to the inorganic matter in 100 parts is 28·70 and 71·30 respectively.

. *Enamel* is deposited on the dentine in vertical layers of horizontal prisms; it is an extremely hard dense substance, resisting the action of the strongest chemical reagents. Of all animal substances it contains by far the smallest quantity of organic matter, as the following table shows.

Composition of enamel in 100 parts.

Organic matter and water	3·59
Calcium phosphate & calcium fluoride	86·91
Magnesium phosphate	1·5
Calcium carbonate	8·0

Cement is a thin layer of bony matter which

connects the tooth fang with the alveolar process, its chemical composition is the same as ordinary bone.

MUSCULAR TISSUE.

COMPOSITION OF MUSCULAR TISSUE.   (Kühne.)

Water	74·0	80·
Solids	26·0	20·
Albuminous substances inso- luble in water, as myosin, sarcolemma, nuclei, etc.	15·4	17·7
Soda Albuminate . . .	2·2	3·0
Gelatin	0·6	1·9
Kreatin	0·07	0·14
Fat	1·5	2·30
Lactic Acid	1·5	2·30
Phosphoric Acid . . .	0·66	0·70
Potash	0·50	0·54
Soda	0·07	0·09
Sodium Chloride . . .	0·04	0·09
Lime	0.02	0·03
Magnesia	0·04	0·05

The tissue of voluntary muscle consists of a number of delicate tubes or fibrillæ, formed by a sheath of *sarcolemma* containing a semi-fluid *plasma*, with nuclei mixed up with elements of connective, adipose, vascular, and nervous tissues.

THE SARCOLEMMA resembles elastic tissue in its chemical characters: it does not yield gelatin by boiling, nor is its elasticity affected by the action o acids and alkalis.   The gelatin which is obtained from muscular tissue is therefore not derived from the sarcolemma, but from the connective tissue which binds the muscular fibrils together.

THE MUSCULAR PLASMA is obtained by injecting the muscles of a freshly killed animal with a 1 per cent solution of sodium chloride; and when the blood is thoroughly washed out, the muscles are cut up into minute fragments, frozen, and mixed with four times their volume of snow containing 1 per cent. of sodium chloride; at 0° C. the mass becomes liquid and must then be filtered rapidly, the filtrate at ordinary temperatures separating into

    (a) *muscle clot*
    (b) *muscle serum.*

MUSCLE CLOT consists of myosin, with which is mixed a colouring matter identical with hæmoglobin, called myochrome (Thudichum).   Formerly syntonin was considered as a chief constituent of the clot, but the researches of Kühne seem to show that syntonin is an artificial product, formed by the action of an acid on myosin and albumin.

MUSCLE SERUM contains albumin of which Kühne recognizes three varieties; viz., one coagulating at a temperature of 30° C. and usually present in serum with a strongly acid reaction; a second which coagulates at 45° C.; and a third, the most abundant, which coagulates at 75° C.   According to Brücke and Piotrowski, in addition to these albumins there is an albuminous ferment which resembles pepsin.

The serum also holds in solution certain extractives as, kreatin, xanthin, hypoxanthin, uric acid, sugar, glycogen, sarcolactic, acetic, formic, and butyric acids; these can be obtained for examination by exhausting the dried residue of muscle serum with alcohol and boiling water.

By evaporation and incineration the serum yields about 2 per cent. of saline residue; in which the potash salts and phosphates greatly exceed the sodium salts and chlorides, as will be seen in the subjoined table.

COMPOSITION OF THE ASH OF MUSCULAR TISSUE.

Sodium chloride . . .	11·5
Sodium sulphate . . .	1·5
Sodium phosphate . . .	11·
Potassium phosphate . .	58·
Earthy and ferric phosphates	18·
	100·0

Muscle serum agitated with ether and the etherial solution evaporated yields on an average 1·5 per cent. of fat; but this quantity is subject to great variations in the human subject, depending very much on the health and activity of the muscles in the body. In fatty degeneration olein seems to replace the more solid constituents of fat, stearin and palmitin.

Living, and to a certain extent dead, muscular tissue absorbs oxygen and eliminates carbonic acid. That this is a vital rather than a chemical process is shown by the fact that the power of absorption diminishes as the temperature rises,

which would not be the case if the process were
a chemical one.

The reaction of voluntary muscular fibre during
life, and at rest, is neutral; when contracted, or
after death it becomes acid, (from the formation
of lactic acid), and remains so till decomposition
renders it alkaline.

Muscular tissue possesses the power of contraction, that is of shortening in length and increasing in its other dimensions. This power, which
is due to the "irritability" of its plasma, is maintained in the living muscle by the circulation of
oxygenated blood, which not only supplies the
oxygen and nutritive materials requisite for the
performance of its functions, but also removes the
products of its disintegration. Brown-Sequard
has illustrated this dependence of contractibility
on the circulation of arterial blood, by causing
fresh human defibrinated blood, which had been
exposed to the air and acquired an arterial tint, to
be injected into the vessels of muscles which had
lost their irritability. Ten minutes after the injection,
the muscles regained the power of contracting
and retained it for two hours, after which the irritability passed off and was succeeded by the "rigor
mortis." The injected blood was returned from the
vessels in a venous condition having lost its arterial
tint.

A certain degree of fluidity is necessary for the
maintenance of muscular irritability; thus, Dr.
Preyer has shown that muscles rendered rigid by
cold, regain their contractility on the injection of

dilute saline solutions of sodium chloride and nitrate.

Oxygen, dilute acids, and alkalis act as powerful
stimulants, and increase the contractility of the
muscular fibres; on the other hand carbonic acid
gas, carbonic oxide, hydrogen, and sulphurous
acid gas diminish it. Woorara, conia, and the
iodides and sulphates of methyl, strychnia, and
brucia, destroy the irritability by their action
through the nervous system; whilst upas antiar
and potassium sulphocyanate destroy it by their
action on the muscle itself.

"Rigor mortis" or cadaveric rigidity was formerly supposed to be due to a spontaneous contraction of the muscles after death, but Kühne has
shown that this phenomenon is produced by the
separation and *coagulation* of myosin; the rigidity lasts till decomposition commences, and the
coagulated material becomes softened and disintegrated.

Muscular tissue, unless there is some degree of
activity, soon loses its power of contracting, and
undergoes fatty degeneration; this is the condition
of muscles in limbs that have long been disused.

Very little is known concerning the decomposi-
tions that occur in muscular tissue as the result of
its activity; it is, probable however that the nitrogenous materials break up into carbonic acid,
lactic acid (to which the acid reaction of active
muscle is due, as well as from the acid phosphates,)
sugar, and fatty matters; and also into kreatin, xanthin, hypoxanthin, etc. This view is

strengthened by the fact that muscle in a state of contraction, *i.e.* in a state of activity, gives off more carbonic acid, acquires an acid reaction, and contains more kreatin, sugar, etc.; than muscle in its elongated condition, *i.e.* at rest.

The force exerted by the muscles is undoubtedly derived from oxidation. Formerly it was supposed that the combustion of the nitrogenous principles was the sole source of muscular force and that urea was the measure of muscular work. This view is now discredited, as urea does not represent an amount of combustion sufficient to account for the whole of the force produced; and as muscular exertion is always attended with increased elimination of carbonic acid and water, which represent the decomposition of the fats and carbo-hydrates, it therefore follows that these bodies furnish a very large proportion of muscular force. Moreover, although great exertion can be borne for some days on a diet free from nitrogen, speedy fatigue and weakness follows the withdrawal of the non-nitrogenous.

Professor Frankland has determined by direct experiment the force values of albumin, fat, and farinaceous matters, when consumed in the body, in kilogrammeters; thus,

1 grm. of Dry Albumin = 1781 kilogrammeters.
1   ,,   Fat of Beef   = 3841   ,,
1   ,,   Starch   = 1627   ,,

If therefore 120 grms. of dry albumin, 90 grms. of fat, and 330 grms. of farinaceous compounds be consumed, 1,096,320 kilogrammeters will be obtained ; which represents a force sufficient to raise the body through one mile in vertical height, or equivalent to walking 20·7 miles per diem.

Nervous Tissue.

Nervous tissue consists of two elements, nerve fibre, and nerve vesicle.

Nerve fibres are found in the nerves and in the *white matter* of the brain and spinal cord. During life they appear as minute, homogeneous, cylindrical, oily-looking filaments of $\frac{1}{1800}$ to $\frac{1}{20000}$ inch in diameter. After death a separation of their contents takes place, and then can be distinguished an outer membrane of fine elastic tissue, *neurilemma,* which does not yield gelatin by boiling, and the elasticity of which is not impaired by the action of acids or alkalis. This investing membrane forms a tube containing a semi-solid fatty matter, *the medullary substance,* soluble in ether, which refracts light strongly. and gives the fibre its dark outline under the microscope ; it is composed of a mixture of fat and albumin the latter probably allied to alkali-albumen or casein. In the centre of the fibre, surrounded by the medullary substance, is the *axis cylinder,* a solid filament insoluble in ether, and receiving a pink stain from a solution of carmine ; it is composed almost entirely of albumin allied to fibrin and myosin, containing but little fat.

The reaction of the nerve fibres during life and in a state of inaction is neutral ; on the application of a stimulus they become acid, and are always so after death, till decomposition sets in.

Nerve vesicles which form the *grey matter* of the brain and spinal cord consist of minute cells of variable form and size, formed by a sheath of neurilemma, containing a mixture of fatty and albuminous substances, with a nucleus and a nucleolus and some granular matter.

The vesicular matter contains more water and less fat and inorganic residue than the white or fibrous matter. The specific gravity of grey matter is 1·034, that of white 1·041.

An aggregation of white with grey matter, forms the brain and spinal cord, and other ganglionic centres, the chemical composition of which may be stated as follows:

Analysis of cerebral matter in 100 parts.

Water	80
Fats	5
Albumins	7
Extractives and Salts	8

WATER. The amount of water varies considerably in different parts of the brain, thus the corpus callosum contains 70 per cent. whilst the cortical substance of the hemispheres contains 84. The spinal cord also contains less water than the brain; in the nerve centres of the fœtus, the child, and the old man there is more water than in those of adult life. The quantity of water is increased in fatty degeneration of these centres.

THE FATTY MATTERS appear to be mixtures of lecithin, oleophosphoric acid, and cerebrin uniting in different proportions to form protagon, and cerebric acid (refer to Lecithin, Cerebrin, Oleophosphoric acid). Glycerophosphoric acid is never found in a free state in fresh and healthy brain, but only in disease or when decomposition has set in (see Lecithin). Cholesterin forms 20 per cent. of the fatty matters of the brain.

THE ALBUMINOUS SUBSTANCES have not yet been satisfactorily determined. The albumin of the axis cylinder is supposed to resemble fibrin and myosin, it is however insoluble in a solution of potassium nitrate and in dilute acids. The albumin of the medullary substance is soluble in dilute acids and resembles casein.

EXTRACTIVES. The aqueous and alcoholic solutions yield elasticin, kreatin, leucin, xanthin, hypoxanthin, lactic and uric acids. .

THE SALTS obtained by incineration amount to 2 per cent., of which the phosphates form by far the largest proportion.

Composition of brain ash in 100 parts.

Potassium phosphate	55·2
Sodium phosphate	22·9
Ferric phosphate	1·2
Calcium phosphate	1·6
Magnesium phosphate	3·4
Sodium chloride	4·7
Potassium sulphate	1·6
Free phosphoric acid	9·0
Silica	0 4

THE DIGESTIVE FLUIDS.

CHAPTER VIII.

THE SALIVA.

THE Saliva is a secretion furnished by the parotid, submaxillary, and sublingual glands; from which it can be obtained by introducing canulas into their respective ducts.

In the mouth it becomes mixed with the oral mucus.

Thus mixed, it is a turbid, viscous fluid, with an alkaline reaction, specific gravity 1·018-1·025 ; it possesses the power of converting starch into glucose which is due to ptyalin.

The fluid secreted by the submaxillary gland, differs from that secreted by the parotid, and sublingual glands, in containing mucin. Irritation of the nerves, which supply the submaxillary gland, as the chorda tympani, the branches of the sympathetic, and the submaxillary ganglion, cause the saliva to become ropy and tenacious, and the alkaline reaction more marked. On the other hand in paralysis of these nerves, the secretion becomes thin and watery.

ANALYSIS OF MIXED SALIVA.

Water 994
Solids 6
Ptyalin . . . . 1·4
Mucus . . . . . 1·5
Fat . . . . . 1·3
Potassium Sulphocyanide . ·10
Salts . . . . . 1·7

PTYALIN is obtained by precipitating fresh saliva with dilute phosphoric acid and lime water; filtering off the precipitate and dissolving it in distilled water; from the clear solution so obtained, the ptyalin is precipitated by alcohol.

Ptyalin is a ferment, though whether it is an albuminoid is doubtful, as it does not give the xanthoproteic reaction with nitric acid; it is soluble in water, does not coagulate when heated; it converts starch into glucose, but this action is does not take place if strong acids or alkalis are added, or if the temperature is raised above 60° C.

POTASSIUM SULPHOCYANATE. (KCNS). The quantity of this substance present in saliva is subject to great variation. Its presence in saliva is shown by the cherry-red colour it gives with ferric chloride, *and which disappears on the addition of mercuric chloride.*◦

The use of this substance has not been satisfactorily explained, Kletzinsky suggests that it prevents the formation of fungi between the teeth.

---

* (This distinguishes it from Meconic acid.)

THE SALTS. The ash of saliva consists principally of the alkaline carbonates, and calcium carbonate ; the latter is frequently deposited on the teeth as tartar.

ORAL DIGESTION. In the mouth the food is thoroughly broken up, triturated, and mixed with the saliva; the fluidity of the secretion separates the particles of the food and allows the digestive fluids to act more freely on them; and its viscidity greatly assists the act of deglutition. The conversion of the starchy matters of the food commences in the mouth, but the change is only partially effected at this stage of the digestive process.

GASTRIC JUICE.

Gastric juice is an acid, glairy, amber-coloured fluid, secreted by the peptic gastric glands.

The best way of obtaining it for examination, is to introduce a small quantity of boiled muscular tissue, free from fat, into the stomach of a dog which has been kept fasting some hours, through a permanent gastric fistula, to collect the juice as it flows, by means of a canula inserted into the fistula, and afterwards to filter the fluid to remove any accidental impurities. The clear gastric juice obtained at this early stage of digestion usually has the specific gravity of 1·010, and about the following composition.

Water	975·00
Organic matter with pepsin	15·00
Free acid	4·78

Sodium chloride   1·70
Potassium chloride   1·08
Calcium chloride   0·20
Ammonium chloride   0·65
Calcium phosphate   1·48
Magnesium phosphate   0·06
Ferric phosphate   0·05

The active property of the gastric juice is due to the principle *pepsin*, and to the *free acid* it contains.

PEPSIN is best prepared, by separating the mucous membrane from the muscular tissue of the stomach of a recently killed animal, rejecting the pyloric extremity, and macerating it in dilute phosphoric acid, till it is completely dissolved. The acid solution is then precipitated with lime water, the precipitate removed by filtration, washed and redissolved in dilute hydrochloric acid. To this acid solution, a saturated solution of cholesterin in 1 part of ether and 4 of alcohol, is added gradually by means of a long filter, passing down to the bottom of the vessel, and the whole mixture well agitated. The cholesterin immediately separates and rises to the surface, bringing the pepsin with it; this mixture of cholesterin and pepsin is then separated by filtration, and the cholesterin removed by repeated agitation with ether.

Thus obtained, pepsin is a greyish-white powder, insoluble in water, alcohol, and ether, very soluble in dilute acids. Its solutions are precipitated and coagulated by heat and alcohol, also by saturated solutions of sodium chloride and some other salts.

Heated with strong nitric acid pepsin does not give the xantho-proteic reaction; hence it would appear that pepsin is not an albuminoid substance as has hitherto been supposed.

Bile added to an acid solution of pepsin is at once precipitated, the whole of the biliary colouring matter being thrown down; the acid reaction is not lost, though the digestive powers of the solution are destroyed.

The characteristic property of pepsin is the power it has, in conjunction with dilute acids, of dissolving albuminous substances and converting them into peptones. The process goes on most rapidly at a temperature of 50° C. (Von Wittich.) Neutralization of the acid fluid likewise arrests its action.

THE FREE ACID. There can be no doubt that hydrochloric acid, and probably lactic acid, are present in a free state in the gastric juice; but considerable difference of opinion exists as to the formation and part played by these acids respectively. Most chemists consider that hydrochloric acid plays the important part in gastric digestion, and that lactic acid, if it has any influence, is of secondary importance. Indeed, some chemists have questioned if lactic acid is present at all in the healthy gastric secretion, and maintain that if it be found it is due to some morbid condition of the organ. The hydrochloric acid is evidently furnished by the decomposition of the alkaline chlorides; but how that decomposition is effected there is no satisfactory evidence to show.

The best degree of concentration of hydrochloric acid for artificial digestive purposes, is 1 per cent. : a high degree of concentration arresting the digestive process.

PEPTONES. These bodies are formed by the action of the gastric juice, or by weak acid solutions of pepsin, on albuminous substances, at a temperature not exceeding 50° C. They are white, amorphous substances, having an acid reaction on litmus paper, soluble in water, insoluble in alcohol; they diffuse readily through porous membranes ; and their solutions turn the plane of polarized light to the left. They are not coagulated by heat, give no precipitate with alcohol, the mineral acids, or potassium ferrocyanide ; they are precipitated by tannic acid, and mercuric chloride ; heated with strong nitric acid they give the xantho-proteic reaction. The peptones mixed with gastric juice interfere with Trommer's test for grape sugar ; and the reaction between starch and an aqueous solution of iodine does not develope the characteristic blue colour, nor is bile precipitated by the gastric juice, if peptones are present.

Several varieties of peptones have been described by Meissner, as,

Parapeptone
Metapeptone
Peptones a, b, and c
Dyspeptone.

They may be obtained for examination as follows. Digest some finely minced rump-steak free from fat, with an acid (1 per cent.) mixture of pepsin,

at a temperature of 50° C., till the muscular fibre
is completely dissolved; then carefully neutralize
the turbid acid fluid with a few drops of sodium
carbonate solution; the precipitate which forms
is parapeptone. This substance is insoluble in
water, soluble in dilute acids, from which solutions
it is precipitated by a mixture of alcohol and ether,
by tannic acid, and mercuric chloride; its acetic
acid solution is precipitated by potassium ferrocyanide. Parapeptone is now generally regarded
as idehtical with syntonin, or acid-albumin, and to
be an intermediate product before conversion into
the other peptones.

The parapeptone being removed by filtration,
the neutralized filtrate is again acidified when
another precipitate, metapeptone, is thrown down;
this is removed by filtration and the *a*, *b*, *c*, peptones remain in the filtrate; these are distinguished
from each other; thus,

(*a.*) Is precipitated by nitric acid; and by
potassium ferrocyanide with a little acetic acid.

(*b.*) Is precipitated by potassium ferrocyanide and much acetic acid; but not by nitric acid.

(*c.*) Is not precipitated by either nitric acid,
nor by potassium ferrocyanide with acetic acid.

The albuminous residue which resists digestion
is called Dyspeptone.

GASTRIC DIGESTION. In the stomach the solution of
the albuminous elements is effected by the action of
the gastric juice, which converts them into diffusible peptones. The gastric juice also dissolves the

gelatinous principles, most probably by converting them into soluble sugar of gelatin.

Most physiologists consider that the oleaginous principles are not acted upon in the stomach, and pass unchanged into the intestines. Dr. Marcet however states, that the neutral fats are resolved in the stomach into fatty acids and glycerin ; and in that condition pass into the intestine, where the fatty acids are saponified by the alkalis of the bile and pancreatic juice.

It is a question too whether the action of the saliva on the starch is continued in the stomach or not. Dr. Brown-Sequard states that the gastric juice does not impede the conversion ; on the other hand Dr. Dalton has convinced himself by frequent experiments, that no further action takes place in the stomach. The latter observer introduced into the gastric fistula of a dog a mixture of starch and meat ; after twenty minutes, the starch was easily recognizable by its reaction with iodine ; and at the end of one hour it had disappeared, but no sugar could at any time be detected. He also found if two mixtures were made, one of starch and saliva, the other of starch, saliva and gastric juice, and both mixtures kept at a temperature of 36° C. for a few minutes, sugar would be found in the first mixture, whilst the second remained unaltered. Therefore Dr. Dalton considers that the final conversion of the starchy matters of the food takes place in the duodenum and other parts of the small intestine, under the influence of the pancreatic and intestinal secretions.

THE BILE.

The Bile is the secretion furnished by the liver, it is a brownish yellow fluid, of a viscid, and mucilaginous nature; its reaction is usually neutral, sometimes alkaline. When shaken it froths and forms a persistent lather; it does not coagulate when heated. Owing to the quantity of mucus it contains, it easily undergoes decomposition, but if the mucus be removed, it will keep fresh for a considerable time. Sp. grav. 1·018-1·020.

Bile is precipitated by the addition of fresh gastric juice; this precipitation does not take place however if the gastric juice holds any peptones in solution.

Bile has no action on the albuminoid and starchy matters of the food.

Bile, agitated with any of the neutral fats, causes them to become subdivided into minute globules, surrounded with a layer of soap, and containing in their interior free fatty acids. Oil, to which a few drops of bile have been added, passes readily through animal membranes under the slight pressure of 0·068 to 0·132 inches of mercury; whereas oil by itself will not pass through unless considerable force be employed.

When acted upon by concentrated nitric acid, bile gives a play of colours, passing from green, blue, violet and red to dirty yellow; this is the test for the bile pigments.

If a solution containing bile be dropped into a

mixture of sulphuric acid and sugar,° a purple red colouration will be formed at the junction of the two liquids; this reaction is known as Pettenkofer's test for bile acids.

There can be little doubt that the principal constituents of the bile, as the conjugated acids and pigments, are separated in the liver, and do not exist preformed in the blood. That this is the case, is shown by the fact that the blood of animals whose livers have been removed contains no traces of these substances, which it certainly would do if they were merely eliminated by this organ. Again, the blood of the hepatic artery and portal vein has been repeatedly examined, without discovering the slightest trace of either bile acid or bile pigment.

From what materials of the blood the bile is formed, is still a matter of conjecture; but the view generally adopted is this: that the circulating albumin, *i.e.* the albumin which has played its part in the economy and is now unfit for use, is broken up in the liver into 1. An amyloid substance or glycogen. 2. Certain fatty acids, and 3. Certain nitrogenous substances, as urea, glycocin, taurin, uric acid, and pigmentary matters.

Of these substances, the glycogen is, most probably, converted into sugar, and in this form passes into the general circulation, and together with the

* A grain of sugar boiled with a drop or two of dilute sulphuric acid at the bottom of a test tube and concentrated sulphuric acid added, is the most convenient way of preparing the test solution.

fatty acids is oxidized to carbonic acid and water, and in this form eliminated by the lungs. The urea and uric acid likewise enter the circulation and are eliminated by the kidneys; whilst the taurin and glycocin united to the cholic acid form the conjugated bile acids; and these together with the pigmentary matters, cholesterin, and other fatty bodies, make up the bile which passes off by the common bile duct into the intestinal canal. What alterations it there undergoes we do not precisely know, but it seems probable that the glycocholate and tauro-cholate of soda, by their decomposition furnish the free alkali necessary for the saponification of the fatty matters; and the glycocin and taurin thus set free are re-absorbed by the intestine; the cholic acid being at the same time decomposed into fatty acids of a simpler constitution.

That some of the biliary substances are reabsorbed, has been shown by Bidder and Schmidt, who by analyzing the fæces of dogs, passed during a period of five days, found the quantity of sulphur contained in them was only $\frac{1}{15}$ of that passed originally into the intestine with the bile.

When any obstruction is offered to the onward passage of the bile into the intestine, a yellow tinging of the skin, as well as of the other tissues and fluids, takes place, giving rise to the phenomenon of jaundice. In these cases the bile is taken up by the hepatic vein and carried into the circulation, and both the bile acids and bile pigments are found in the urine. In some cases, however, jaundice occurs when there is no obstruction to the onward passage of bile into the bowels, and Kühne considers that in these cases there is no re-absorption of the biliary matters, but the jaundice is produced by the colouring matter of the blood, from the introduction of certain septic and poisonous substances into the blood, which dissolves the blood corpuscles, and converts their freed colouring matter into bile colouring matter; in these cases the bile acids are not present in the urine.

Frerichs stated that when the biliary acids were injected into the blood they were transformed into bile pigments, and that the urine became darkly coloured in consequence. Kühne, on the other hand, has shown that the bile acids when injected into the blood are not decomposed, but pass unchanged out of the system into the urine; Kühne's views have received the support of Hoppe Seyler, Virchow, and Neukomm.

Composition of Human Bile.

Water	860
Solids	140
Glycocholate of soda } Taurocholate of soda }	90·8
Fat	9·2
Cholesterin . . .	2·6
Mucus	1·4
Pigment and Extractive .	28·
Salts	8·

ANALYSIS. *Method of proceeding.* Divide the bile for examination into 3 portions.

(a) Evaporate over a water bath to determine the amount of water and total solids; incinerate the residue over a Bunsen lamp to get the quantity of fixed inorganic salts.

(b) Remove the mucus by precipitation with acetic acid, filter, evaporate the clarified bile to a syrupy consistence, and mix with animal charcoal; introduce the mass whilst still hot into a flask containing alcohol, and let it digest some days; this forms the *Alcoholic Extract of Bile.*

(c) Free from mucus, and evaporate at a gentle heat to near dryness, this forms *Inspissated Bile.*

WATER. The proportion of water to the solid constituents of the bile is extremely variable, being from 85 to 95 per cent. In cholera, and febrile affections, the quantity of water is diminished; calomel, and other cholagogues, increase it. Dr. Dalton has shown that, the quantity of bile secreted is increased after food; and the proportion of solid matter to the water is gradually increased during digestion; thus,

Time after eating.	Quantity of Fluid.	Solids.
Immediately	640	33
1 hour	1990	105
3 hours	780	60
6 ,,	750	73
9 ,,	860	78
12 ,,	325	23
21 ,,	384	11
24 ,,	164	$9\frac{1}{2}$

From this table it will be seen that the greatest quantity of fluid is secreted one hour after eating, and that then the solids constitute about 5 per cent. The proportion of solids to the water gradually rises till the 6th hour, when they form about 9½ per cent.; after this time they again decrease; at the end of the 9th hour, they are 9 per cent.; at the end of the 12th, only 7 per cent.; and at the end of 24 hours, are again 5 per cent.

THE BILE ACIDS are formed by the conjugation of taurin and glycocin respectively, with cholic acid; and in the bile they are always found as sodium salts.

GLYCOCHOLIC ACID.

Preparation. Some of the *alcoholic bile extract* is filtered; and to the clear solution an excess of ether is added, when a pulverulent white deposit will be thrown down. This is filtered off, dissolved in water, and the aqueous solution precipitated with *neutral* lead acetate. This precipitate is filtered off, washed, and dissolved in alcohol, the lead removed by precipitation, with sulphydric acid, and filtered. The clear filtrate, diluted with water, on standing will deposit crystals of glycocholic acid.

Properties. The crystals are long, delicate, colourless needles, which lose weight when heated to 100° C.; they are slightly soluble in cold water and ether, very soluble in alcohol and boiling water; their solutions turn the plane of polarized light to the right; they are precipitated by the *neutral* lead acetate. Heated with an excess of

baryta water they are decomposed, forming barium cholate and glycocin. They give a deep
purple colour with concentrated sulphuric acid
and cane sugar (Pettenkofer's test).

TAURO-CHOLIC ACID.

Preparation. The mother liquor, left after the
precipitation of the glycocholic acid, is to be precipitated by *basic* lead acetate; and the precipitate
treated in the same way as directed for glycocholic
acid.

Properties. Taurocholic acid never occurs in a
crystalline form, but appears as an oily resinous
fluid, of tawny colour, very soluble in alcohol and
ether, and has a strong acid reaction; its aqueous
solution, on heating, is readily decomposed into
taurin and cholic acid; it turns the plane of polarized light to the right, and gives the purple reaction with sulphuric acid and cane sugar.

FATTY MATTERS. Dried inspissated bile is shaken
up with ether, and the etherial solution evaporated;
when a mixture of neutral fats, cholesterin, and
lecithin will be left as a residue.

EXTRACTIVES. The alcoholic extract contains an
energetic base, choline or neurine (see p.) and
small quantities of xanthin, hypoxanthin, tyrosin,
and leucin; in Bright's disease urea is also found.

SALTS. The ash of bile yields about 27 per cent
of sodium chloride, 28 per cent of alkaline phosphates, the remainder being made up of alkaline
carbonates, earthy phosphates, and small quantities of iron and silica.

PIGMENTS. The bile pigments have been thoroughly investigated by Thudichum and Städeler; they consist of an orange-red pigment, bilirubin, a brown pigment, bilifuscine, and two green pigments, biliverdin, and biliprasin, and an insoluble residue, bilihumin.

BILIRUBIN	$C_{16}H_{18}N_2O_3$
BILIFUSCINE	$C_{16}H_{20}N_2O_4$
BILIVERDIN	$C_{16}H_{20}N_2O_5$
BILIPRASIN	$C_{16}H_{22}N_2O_6$
BILIHUMIN	

BILIRUBIN.

Preparation. Extract some inspissated bile, or better still if at hand, some crushed gall-stones, successively with water, alcohol, dilute hydrochloric acid, boiling alcohol, and ether ; the dried residue is then boiled with pure chloroform ; the chloroform extract is distilled to near dryness, and several volumes of alcohol added, which throws down the bilirubin.

Properties. Bilirubin is an orange red powder, mixed with a few bluish brown crystals ; insoluble in water and ether; freely soluble in chloroform, bisulphide of carbon, turpentine, and benzol; only slightly soluble in alcohol. The chloroformic solution is orange red, but loses its colour on the addition of an alkaline solution. An ammoniated solution of bilirubin gives a rust-coloured precipitate, insoluble in chloroform, in calcium, and barium chloride solutions, and with lead acetate. On adding fuming nitric acid to a solution of bilirubin, a play of colours, passing from green, blue, violet, and red to a dirty yellow takes place. (Test for bile pigments). Concentrated sulphuric acid dissolves bilirubin, forming a brown solution.

Bilirubin is considered by some chemists to be identical with hæmatoidin, but the researches of Koln, Robin and Verdeil, show that this is not the case. (See hæmatoidin.)

BILIFUSCINE.

Preparation. A chloroformic solution of bile or crushed gall-stones is evaporated to dryness, and the residue dissolved in alcohol, and evaporated ; the residue shaken up with ether and chloroform, and the insoluble portion dissolved again in alcohol. The alcoholic solution on evaporation leaves the bilifuscine as a dark brown residue.

Properties. It is insoluble in water, ether, and chloroform, soluble in alcohol and alkalis; hydrochloric acid throws it down from its solution in brown flocks. Its ammoniated solution gives a brown precipitate with calcium chloride. Bilifuscin according to Städeler consists of bilirubin, plus one atom of water; thus

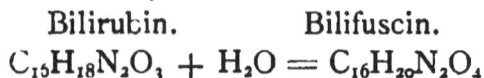

Bilirubin.   Bilifuscin.

$$C_{16}H_{18}N_2O_3 + H_2O = C_{16}H_{20}N_2O_4$$

BILIVERDIN.

Preparation. Is formed by passing a current of air through an alkaline solution of bilirubin.

Properties. The green solution thus obtained deposits green flocks on the addition of hydrochloric acid, which become black when dried; the flocks are soluble in alcohol, benzol, and carbon bisulphide, insoluble in water, ether, and chloroform. An ammoniated solution is not precipitated by calcium or barium chloride; its alcoholic solution gives a dark green precipitate with baryta and lime water, also with silver, lead, and mercury acetates. According to Städeler biliverdin is formed from bilirubin by the addition of one atom of water in the presence of oxygen; thus

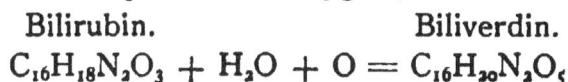

Bilirubin. Biliverdin.

$$C_{16}H_{18}N_2O_3 + H_2O + O = C_{16}H_{20}N_2O_5$$

BILIPRASIN.

Preparation. Inspissated bile or crushed gallstones are treated successively with ether, hot water, chloroform, and dilute hydrochloric acid, and the residue treated with boiling chloroform; the portion left undissolved is dissolved in alcohol, which on evaporation leaves the biliprasin as a brittle, shining mass of dark green colour.

Properties. It is soluble in alcohol and the caustic alkalis; insoluble in ether, and chloroform; the green alcoholic solution becomes brown on the addition af alkalis, and again becomes green on the addition of acids.

BILIHUMIN.

Is the insoluble residue left after the bile or biliary calculi have been exhausted by ether, water, weak acids, chloroform, and alcohol.

PANCREATIC FLUID.

The pancreatic fluid may be obtained for examination by opening the abdomen of a dog, and drawing down the duodenum and separating the lower and larger pancreatic duct, and passing a canula into it; the duodenum is then returned, and the wound closed by a ligature, the canula being left hanging out.

It is a clear, viscid fluid, free from smell, having an alkaline reaction; specific gravity 1·010—1·013; heated to 54° C. it becomes turbid, and coagulates in white flakes at 60° C. It is precipitated by the addition of alcohol and tannic acid. It converts boiled starch into glucose, and the addition of chlorine water gives it a rose-coloured tint.

COMPOSITION OF PANCREATIC FLUID.

Water 900.76
Organic matter with pancreatin* 90·38
Sodium chloride . . . . 7·36
Free sodium . . . 0·32
Sodium Phosphate . . . 0·45
Sodium Sulphate . . 0·10
Potassium Sulphate . . . 0·02
Combinations of—
  Lime . . . . . 0·54
  Magnesia . . . . 0·05
  Iron . . . . . 0·02

* Pancreatic juice often contains small quantities of leucin and tyrosin.

PANCREATIN is obtained by rubbing down the pancreas of a freshly killed animal, in full digestion, with pounded glass, from which an aqueous solution is made, and from which the pancreatin may be precipitated by alcohol.

Pancreatin is an albuminoid substance which rapidly decomposes; it is soluble in water, and its solutions are coagulated by heat, by mineral acids, and by magnesium sulphate.

Pancreatic juice converts starch into glucose, thus completing the metamorphosis commenced by the saliva. It also assists in the digestion of fatty matters by converting them into an emulsion; this emulsion, according to Dr. Marcet, is formed of minute globules, each surrounded with a layer of soap, and containing fatty acids in its interior. Pancreatic fluid also has the power of converting albuminous substances into peptones, which differ from the gastric peptones by being precipitated by acids and acid salts.

According to Kühne the prolonged action of pancreatic juice on newly formed peptones leads to the formation of leucin and tyrosin.

THE INTESTINAL JUICE.

Owing to the great difficulty in isolating the intestinal secretion, very little is positively known regarding it. According to Thierry, the glands of Brunner, and the follicles of Lieberkühn, secrete a yellowish, viscid fluid of alkaline reaction,

L

having a specific gravity of 1·011, and containing
about 2·5 per cent. of solids. He obtained it by
isolating the loop of intestine with a ligature,
below the pancreatic and biliary ducts, and collecting the intestinal juice through an artificial fistula,
opening in the abdominal walls. This juice, according to Thierry, has no action on starch, fat,
or albumin; but Bidder and Schmidt believe, that
like the pancreatic juice, it has the power of chang-
ing albuminoids into peptones, starch into glucose,
and that it has an emulsifying action on fat.

INTESTINAL DIGESTION. The food, after it has been
submitted to the action of the gastric juice, passes
out through the pylorus as acid chyme. This
chyme, consists of gastric juice holding the peptones in solution, and the oleaginous, and starchy
principles of the food as yet unchanged. In the
duodenum, the chyme is mingled with two secretions; viz., the bile and the pancreatic fluid.
These fluids decompose the oleaginous matters,
free fatty acids being formed, which uniting with
the alkaline bases of the secretions, become saponified, and in this condition are more readily absorbed by the intestine. Dr. Marcet, however,
states that the decomposition of the fats into fatty
acids takes place in the stomach, and that saponification alone is performed in the intestines. Pancreatic juice completes the conversion of the starch
into sugar, and also has the power of converting
the albuminous principles into peptones.

The action of the intestinal juice on the different
principles has not yet been definitely determined,

it is most probable that it resembles pancreatic fluid in its action on the various principles of the food.

• As the fluid contents of the intestine pass onward, they become more consistent, from the withdrawal of their soluble portions, by the lacteals and blood vessels; till, by the time they reach the large intestine, they contain little else but the insoluble residue of the food, and the altered excremtitious matters of the bile. In the first portion of the large intestine, the re-action of the contents is often extremely acid, from the lactic and butyric acid fermentation of the starchy matters which have escaped conversion into glucose in the small intestine; but in the lower part of the large intestine, the acid reaction is less marked, and the mass acquires the consis-tence of fæces in which form it is cast out of the alimentary canal.

CIRCULATORY FLUIDS.

CHAPTER IX.

BLOOD.

THE blood is an homogeneous looking fluid of alkaline reaction, having a saltish taste, and a faint odour characteristic of the animal from which it is drawn. Its specific gravity varies from 1·050 to 1·060, the average being 1·055. The temperature of the blood varies from 36·5° C. to 37·8° C., according to the part of the body from which it has been obtained; thus, the blood of the hepatic and portal veins has a higher temperature than ordinary venous blood, and the blood of the right ventricle higher than that of the left. When removed from the body, blood undergoes spontaneous coagulation, separating into a semi-solid clot or crassamentum and a liquid portion or serum.

The red colour of arterial, and the purple colour of venous blood is probably due to the chemical alteration of some substance contained in the blood corpuscles. According to the researches of Hoppe Seyler and Professor Stokes, the hæmoglobin, or as the latter calls it cruorin, which constitutes the greater part of the corpuscles, has the power of holding oxygen in loose combination and forming oxy-hæmoglobin (or scarlet cruorin). In this state it possesses a scarlet hue and gives to the spectrum two absorption bands,✪ one close to D in the yellow, the other close to E in the green, a clear space intervening; if some reducing agent which removes the oxygen be added, as ammonium sulphide, or ammoniacal ferrous sulphate to which enough tartaric acid has been added to prevent precipitation, the oxy-hæmoglobin will be reduced and change from a scarlet to a purple tint; at the same time the two absorption bands will disappear and a broad shadow† appear in the intermediate space that previously was clear. If this reduced hæmoglobin (or purple cruorin) be exposed to the air, it again absorbs oxygen, acquires a scarlet tint, and the two absorption bands are restored to the spectrum. These changes are best observed in solutions of $\frac{1}{3}$ of a inch thick and containing $\frac{1}{1000}$ part of hæmoglobin; they are also visible in solu:. tions of $\frac{1}{10000}$.

It thus appears that oxygen on entering the body chemically combines with hæmoglobin, forming oxy-hæmoglobin, which gives the scarlet colour to arterial blood. In the course of the circulation this becomes reduced, and is returned to the lungs as hæmoglobin with the venous blood.

Arterial blood is monochromatic, *i.e.* when viewed in thin layers by transmitted light it still retains its red tint. Venous blood, on the other hand, is dichromatic, *i.e.* the purple colour ac-

* See Fig. 1. Page 160.　　† See Fig. 3. Page 160.

quires a greenish tint when viewed by transmitted light. Arterial blood differs slightly from venous in containing more water and fatty matter.

COMPOSITION OF BLOOD.

Water . . . 795
Solids . . . 205
Fibrin . . . . 2.
Albumin . . . . 70.
Hæmoglobin . . . 120.
Fatty matters . . . 2.
Extractives . . . 3.
Inorganic residue . . 8.

ESTIMATION OF THE WATER AND SALTS. Take a clean porcelain capsule of known weight and place in it a definite quantity of fresh uncoagulated blood, say 50 grms., and evaporate over a water bath till it ceases to lose weight. On weighing, the loss will represent the amount of water withdrawn from 50 grms. of blood; for example, if the united weight of the capsule and 50 grms. of blood is 110 grms. before evaporation, and only 70·5 after, the blood will have lost 39.5 grms. of water; and $\frac{39·5 \times 1000}{50}$ = 790 grms. the quantity of water in 1000 parts.

The residue left in the porcelain capsule, and which by the above weighing was found to weigh 70·5 grms., is introduced into a muffle furnace or over a Bunsen's lamp, and incinerated till all the organic matter disappears; this residue is again weighed, and its weight represents the amount of inorganic residue in 50 grms. of blood. If for example, after incineration the weight of the capsule and residue is 60·4 grms., then 10·1 grms. of organic matter has been burnt off, and 0·4 grms. represents the inorganic residue of 50 grms. of blood, therefore,

$$\frac{0·4 \times 1000}{50} = 8 \text{ grms. the inorganic residue in 1000}$$

parts of blood.

The quantity of water present in blood is subject to great variation. It is increased by the ingestion of fluids and by deprivation of solid food. It is diminished by exercise, and excessive action of the skin and kidneys. The fœtal blood contains less water than the maternal. Loss of blood and abstinence increase the watery elements of the blood. In some diseases, the water is diminished, as in fevers, cholera, diarrhœa, and the like. It is increased, however, in all diseases which diminish the solid constituents of the blood, as in phthisis, anæmia, leucocythæmia, and chlorosis.

The following table gives the composition of the ash of human blood, it shows the great excess of the sodium over the potassium salts.

ANALYSIS OF ASH OF HUMAN BLOOD.

Phosphoric anhydride .	0·1103
Sulphuric anhydride .	0·0358
Chlorine . . .	0·2805
Potash	0·0343
Soda	0·3748
Lime	0·0112
Magnesia . . .	0·0058
Iron oxide . . .	0·0948

Calculated total 0·8640; found 0·8922. (Jarisch.)

ESTIMATION OF THE FATTY MATTERS AND EXTRAC-
TIVES. Evaporate to dryness over a water bath 50
grms. of blood, and when dry break up the residue
carefully in a mortar and exhaust several times with
boiling ether. Evaporate the etherial solution in a
weighed platinum capsule and again weigh. The
increase in weight represents the fatty matter in
50 grms. of blood ; thus, if the weight of the capsule is 14·8 grms., and after the evaporation of the
etherial solution it weighs 14·9 grms., then there
is 0·1 grm. of fatty matter in 50 grms. of blood ;
therefore, $\dfrac{0\cdot1 \times 1000}{50} = 2\cdot$ grms. in 1000 grms.

The fatty matters consist of a mixture of the following substances—

Saponified fats . .	1.5
Phosphorized fats .	·4
Cholesterin . . .	·08
Serolin . . .	·02
	2·00

THE EXTRACTIVES. The residue, after exhaustion
with ether, is boiled in water for about an hour,
and the aqueous solution filtered and evaporated
to dryness. The residue thus obtained is treated
with alcohol till nothing more is taken up by it,
the alcoholic solution is then evaporated and the
residue weighed, which gives the weight of alcoholic extractives. The residue, which is not taken
up by alcohol, is dried and weighed and represents the aqueous extractive matters. But both
the alcoholic and aqueous extractive matters are

associated with traces of inorganic matter, therefore they must be incinerated, and the weight of
the inorganic residue deducted from the weight of
the dry mass previous to ignition; thus the weight
of alcoholic residue previous to ignition is 0·27
grms., and after ignition it weighs 0·20 grms., or

·07 grms. in 50 grms. of blood, or $\dfrac{·07 \times 1000}{50} = 1·4$

in 1000 parts. And similarly, if the aqueous extractive weighs 0·28 grms. before ignition and ·21
grms. after, then the aqueous extractives amount
to ·07 grms. of extractives in 50 grms. of blood, or
$\dfrac{·07 \times 1000}{50} = 1·4$ grms. in 1000.

The *Extractives* consist of Sugar, Urea, Kreatin,
Uric Acid, Lactic Acid, Hippuric Acid, Leucin,
Tyrosin, Hypoxanthin and Xanthin.

ESTIMATION OF FIBRIN. Carefully clean and dry
a bottle capable of holding 8 ounces, provided with
a stopper, and introduce into it several strips of
lead; then weigh the whole, and after recording the
weight, place in it 5 ounces of fresh uncoagulated
blood, and agitate briskly for twenty minutes; at
the end of that time the fibrin will have separated
and attached itself to the fragments of lead. The
bottle is again weighed, to ascertain the exact
quantity of blood employed, and the blood removed; the bottle, with the fragments of lead and
the adhering fibrin, are carefully washed in cold
water till perfectly free from colouring matter,
and then dried over a water bath, and when thoroughly dry again weighed; the increase in

weight of the bottle and lead corresponds to the amount of fibrin removed from the blood.

ESTIMATION OF ALBUMIN. Take 50 C.C. of fresh uncoagulated blood and weigh it, and then set it aside to coagulate; at the end of 24 hours remove the clot, press it and allow it to drain into the serous portion. Collect the serum, weigh, and evaporate it in a weighed porcelain capsule; thoroughly exhaust the residue by boiling successively with ether, alcohol, and water. The residue, after being exhausted, is dried and weighed, the weight representing the albumin present in 50 C.C. of blood, less a small quantity of inorganic residue which must be deducted, as directed in the case of the alcoholic and aqueous extractives. For example, 50 C.C. of blood weighs 54 grms.; the serum after the clot is removed weighs 44 grms., and the weight of the porcelain capsule is 25 grms. Thus the serum and capsule weigh 69 grms.; after evaporation and exhaustion, the capsule and residue weigh 28·5; therefore, 3·5 represents the quantity of albumin together with a small quantity of inorganic matter in 50 C.C. of blood. To deduct this inorganic matter, the albuminous residue is incinerated, and the ash weighed. Suppose the weight of the ash is ·05 grm., then this deducted from 3·5 grms., leaves 3·45 grms. of pure albumin in 50 C.C. of blood; and $\dfrac{3 \cdot 45 \times 1000}{50} = 69 \cdot$ grms. in 1000 C.C. of blood.

BLOOD CORPUSCLES.

The human red blood corpuscles are minute, discoidal bodies, nearly transparent, and of a yellowish colour, varying in diameter from $\frac{1}{3000}$ to $\frac{1}{4000}$ of an inch, the average being put at $\frac{1}{3200}$, and about $\frac{1}{12500}$ of an inch in thickness. In the natural state their surfaces are bi-concave, but if placed in fluids of less density than the serum, they swell up and become bi-convex and globular, and may even burst: on the other hand, fluids of greater density than serum cause them to assume a shrunken, granulated appearance. They contain no nucleus, the bright spot in the centre being due to the unequal refraction of transmitted light. Solutions of potash, ammonia, and acetic acid render them pellucid by setting free their colouring matter, but does not dissolve them.

ESTIMATION OF THE RED CORPUSCLES. Stir a definite quantity of freshly drawn blood with the plume of a feather to remove the fibrin; add a 10 per cent. solution of sodium chloride, and set aside the mixture in a cool place till the corpuscles are deposited. The mass is then to be collected on a filter, and washed with the solution of sodium chloride till it is perfectly freed from serum; the mass is then placed in a weighed capsule and dried, the weight after drying representing the amount of corpuscles present in a certain quantity of blood.

COMPOSITION OF THE RED CORPUSCLES. The red corpuscles are formed of a delicate membrane or *stroma*, which contains the colouring matter, hæmoglobulin, cholesterin, phosphorized fat, paraglobulin, and inorganic salts, chiefly potassium chloride, and sodium phosphate. The *stroma* is the colour-less portion of the living blood corpuscle, it is insoluble in water, and in sodium chloride solutions, but freely soluble in ether, chloroform, caustic soda, ammonia, and in solutions of the bile acids and urea. The stroma appears to combine with the hæmoglobin and so to speak fixes it, but the union is very feeble, and very slight disturbing influences set free the colouring matter. The hæmoglobin in the living blood is combined with an alkali, probably potash, to keep it in solution, as otherwise it is very insoluble and would crystallize out.

Dr. Preyer has calculated that one blood corpuscle contains 0·000,000,000,02 grm. of hæmoglobin!

Blood corpuscles agitated with ether yield cholesterin and phosphorized fatty matter. The quantity of cholesterin in the corpuscles of 100 C.C. of blood is 0·04 to 0·06 grm. (Hoppe Seyler).

HÆMOGLOBIN.

This name was given by Hoppe Seyler in 1864 to the red colouring matter of the blood contained in the corpuscles, and which separates from them in crystalline forms at a temperature of 0° C.

Preparation. Freshly drawn blood is received into a saucer surrounded with ice, de-fibrinated, and a 10 per cent. solution of sodium chloride added. This mixture is allowed to stand some time till the corpuscles are deposited, the supernatant liquid is then decanted off and the mass washed on a filter repeatedly with sodium chloride solution. When the mass is free from serum, it is to be agitated with a mixture of 1 vol. of water and 4 vols. of ether; the water dissolves the hæmoglobin, the ether, the cholesterin, and phosphorized fats.

The red aqueous solution is then filtered, received into a beaker surrounded by ice, and alcohol added till a precipitate begins to appear. The mixture is then set aside for some hours. If the blood used in this process be obtained from the dog, rat, squirrel, or guinea-pig the crystals will be abundant, and well defined, but in the case of the blood of man, ox, sheep, or horse the hæmoglobin is deposited in an amorphous state, the crystalline form being rare, and only obtainable when the process has been carried out below 0° C.

Chemical and physical properties. The crystals of hæmoglobin are formed upon the rhombic system, the forms varying in different animals; thus, in man, though obtained with difficulty, the crystals consist of four-sided prisms with dihedral summits; in the guinea pig the crystals are tetrahedral; in the rat tetrahedral and octohedral; in the dog and cat the crystals are needle shaped terminated by one plane surface; in the squirrel the crystals are hexagonal.

The crystals are soluble in water and in alkaline solutions, but insoluble in alcohol, choloroform, ether, fatty oils, benzole, turpentine, and carbon bisulphide. Hæmoglobin, though a crystalloid, does not diffuse through parchment paper, thus forming an exception to Graham's theory.

Hoppe Seyler gives the composition of dried crystallized hæmoglobin, as,

$$
\begin{array}{llll}
C & . & . & . & . & 54\cdot01 \\
H & . & . & . & . & 7\cdot20 \\
N & . & . & . & . & 16\cdot17 \\
Fe & . & . & . & . & \cdot42 \\
S & . & . & . & . & \cdot72 \\
O & . & . & . & . & 21\cdot48 \\
\hline
& & & & & 100\cdot00
\end{array}
$$

from which the formula $C_{600}H_{960}N_{154}FeS_3O_{179}$ may be adduced ; and

$$
\begin{array}{lll}
12 \times 600. & C = 7200 \\
1 \times 960. & H = 960 \\
14 \times 154. & N = 2156 \\
56 \times 1. & Fe = 56 \\
32 \times 3. & S = 96 \\
16 \times 179. & O = 2864 \\
\hline
& 13332
\end{array}
$$

gives the molecular weight 13332. And as one molecule of hæmoglobin requires three molecules of soda to form non-coagulable combinations, therefore $\frac{13332}{3} = 4444$, or the equivalent weight of hæmoglobin.

The optical properties of Hæmoglobin and its compounds. It has already been stated that hæmoglobin has the power of forming a loose combination with oxygen, and in this state presenting two ab-

sorption bands in the spectrum characteristic of
arterial blood. This and other changes produced
by different reagents will now be more fully described. If we take a concentrated solution of
oxy-hæmoglobin or undiluted blood, and place it '
in a tube in the slit of the spectroscope we find the
whole spectrum is obscured with the exception of
the extreme red rays; on gradual dilution of the
solution the spectrum clears up, allowing light to
pass in the green beyond E to F, and in the blue
towards G in the violet end of the spectrum; at
the same time in the yellow and beginning of
green between D and E, the spectrum is still ob-
scure; as the dilution is continued the dark space
disappears, leaving two absorption bands a and
β with a clear space intervening; the band a,
which is near D, is smaller and darker than b,
which is near E, (see Fig. 1.) Now if we de-
prive the blood or oxy-hæmoglobin of its oxygen
by the action of reducing agents,* being careful at
the same time to exclude the air, we find the two
absorption bands fade away and a single broad
shadow γ takes the place of the previously clear
space; this band represents the spectrum of un-
combined hæmoglobin (see Fig. 3.) In the spectrum of venous blood this single band is always
present, but as venous blood is never quite free
from oxygen there is always some trace of the
spectrum of oxy-hæmoglobin as well.

* Ammonium sulphide, or ammoniacal ferrous sulphate, to
which enough tartaric acid has been added to prevent precipitation.

Absorption Spectra of the Blood and of its Colouring Matter.

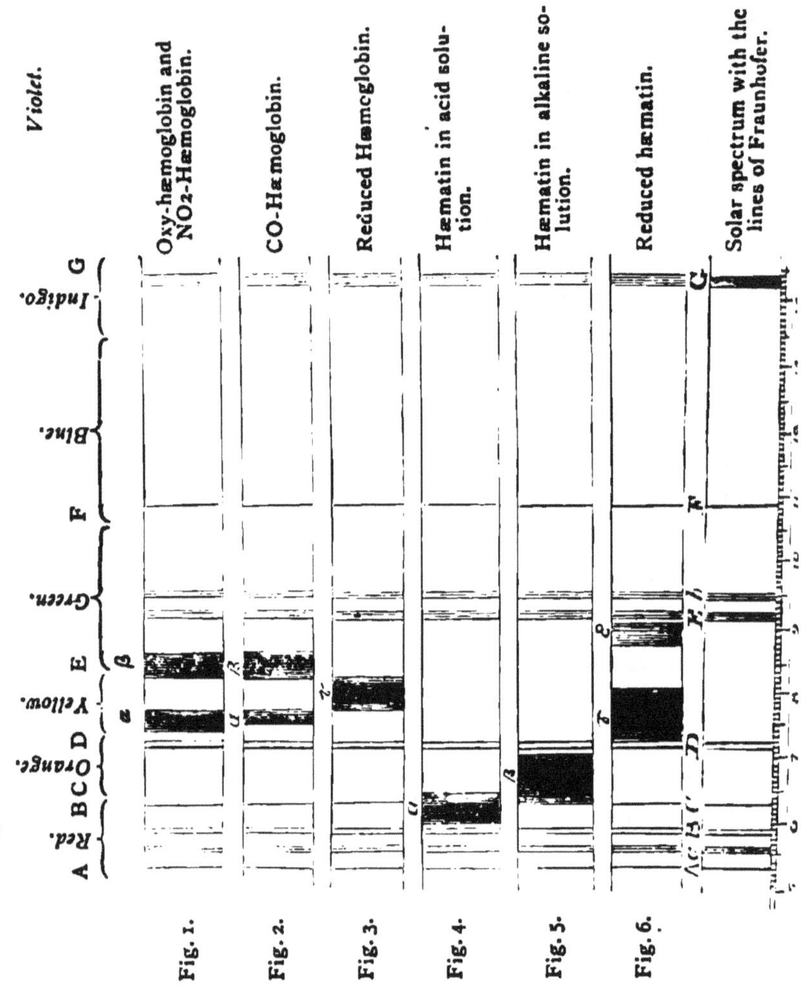

Fig. 1. Oxy-hæmoglobin and NO₂-Hæmoglobin.

Fig. 2. CO-Hæmoglobin.

Fig. 3. Reduced Hæmoglobin.

Fig. 4. Hæmatin in acid solution.

Fig. 5. Hæmatin in alkaline solution.

Fig. 6. Reduced hæmatin.

Solar spectrum with the lines of Fraunhofer.

Carbon monoxide compound. On passing a stream of carbonic oxide through a solution of oxy-hæmoglobin, the oxygen is displaced; at the same time, the solution acquires a dark bluish red tint, but the spectrum is only a little altered; the line *a* being shifted slightly towards E (see Fig. 2.) the absorption bands, however, do not disappear on the addition of reducing agents, and it is by this that the presence of carbon monoxide may be detected in the blood of animals poisoned by it.

Nitrous dioxide compound. A current of nitrous dioxide passed through a solution of reduced hæmoglobin restores the two absorption bands a and β, but these cannot again be made to dissappear by reducing agents. Nitrous dioxide also replaces the oxygen of oxy-hæmoglobin without affecting the two absorption bands, but the entire spectrum becomes darkened.

Potassium and hydrogen cyanide compound. Hæmoglobin combines directly with these substances, and gives with them the two absorption bands as with oxy-hæmoglobin, only the combination appears to be more stable. For solutions of oxyhæmoglobin and prussic acid enclosed in a tube retain their original optical properties much longer than solutions of oxy-hæmoglobin only.

Decomposition of hæmoglobin. Solutions of hæmoglobin readily decompose at temperatures above 0°C., and on the addition of acids, and caustic alkalis, it breaks up into hæmatin and globulin yielding about 4 per cent. of the former to 96 of the latter. With glacial acetic acid and any metallic chloride, it is decomposed into hæmin and globulin.

Hæmatin $C_{68}H_{70}N_8Fe_2O_{10}$. (Hoppe Seyler.) Is best prepared by mixing fresh defibrinated blood with a strong solution of potassium carbonate, till the liquid adhering to the separated coagulum becomes colourless. The coagulum is

M

then dried at 50° C. and digested for some days in absolute alcohol; the alcoholic solution after concentration will deposit rhombic crystals. Hæmatin crystals are of a bluish black colour with a metallic lustre, becoming brown on trituration. They are insoluble in water, alcohol, ether, and chloroform; but soluble in acids and alkalis, the acid alcoholic solutions are monochromatic, having a brown colour; the alkaline solutions are dichromatic and have an olive green colour and dark red in the thicker layers. The crystals support a temperature of 180° C., but above that they carbonize.

When treated with strong sulphuric acid hæmatin is deprived of its iron; this substance to which Hoppe Seyler has assigned the name of *hæmatopopphyrin*, gives to the spectrum a dark band midway between D and E and a narrow band between C and D (nearer D), the spectrum also is deeply shaded between D and E. If a solution of hæmoglobin reduced by hydrogen is decomposed with sulphuric acid, a substance is formed which Hoppe Seyler has named *hæmochromogen*, and which by oxidation yields hæmatin; bilirubin seems closely allied to this substance.

Spectrum of hæmatin. If a little acetic acid be added to a solution of hæmoglobin the two absorption bands *a* and *b* vanish and another absorption band appears, which covers C and extends slightly towards D (see Fig. 4, p. 160); this is the spectrum of hæmatin in an acid solution. If an alkali, such as ammonia, be added, a broad stria appears

which touches C and extends nearly to D; this is
the spectrum of hæmatin in an alkaline solution
(see Fig. 5, p.. 160). If a solution of ferrous sulphate with tartaric acid and ammonia be added,
the two absorption bands disappear, to be replaced by two new ones; viz., a broad band
reaching from D half way to E, and d a narrow
band situate near E (see Fig. 6, p. 160); this is
the spectrum of reduced hæmatin.

Hæmin $C_{68}H_{72}N_2Fe_2O_{10}Cl_2$, or hæmatin hydro-
chloride; if a small quantity of blood is rubbed
up with sodium chloride and boiled for a few
minutes with glacial acetic acid, and the mixture
evaporated to dryness. In the residue mixed with
colourless crystals of sodium chloride, and sodium
acetate, will be found rhombic tablets of hæmin;
which are of a bluish red colour when viewed by
reflected, and brownish red by transmitted light.

The crystals are insoluble in hot and cold water,

in alcohol and ether. Soluble
in alkaline solutions. All acids
with the exception of hydrochloric and acetic acid decompose
them. Heated to 200° C. they

Fig. 8. Hæmin Crystals. undergo decomposition evolving
fumes of prussic acid and leaving a residue of
oxide of iron.

Hæmatoidin $C_{15}H_{18}N_2O_3$, (Robin and Verdeil.)
Under this name Virchow described certain red
crystals found in clots of old extravasations, as in
apoplectic clots, corpora lutea, etc. It has been
considered to be identical with bilirubin, but it

M 2

differs from that substance in chemical composition and in being soluble in ether and insoluble in alkalis. M. Koln has obtained hæmatoidin crystals by rubbing up the corpora lutea with pounded glass and agitating frequently with chloroform; after standing some time the chloroform solution is poured off and allowed to evaporate. The crystals thus obtained are red by transmitted, and green by reflected light; they occur in triangular plates with rounded borders, and are soluble in chloroform, bisulphide of carbon and in ether; they are insoluble in alcohol water and alkalis.

THE DETECTION OF BLOOD STAINS.

To ascertain the nature of the stain, the surface or substance of the material must be scraped or cut up into small fragments and digested in water.

1. Of the reddish fluid thus obtained, a small portion is placed in a deep narrow cell, and examined by a spectroscopic eyepiece with a low power of the microscope. If two absorption bands appear, one close to D and the other close to E with a clear space intervening, the stain is caused by the blood of some red-blooded animal, (see H. C. Sorby, Detection of Blood by its Absorption Spectrum, *Chemical News*, 1865.)

2. A few drops of the fluid are then mixed with a little glacial acetic acid and a small quantity of sodium chloride, evaporated to dryness at 40° to 60° C., and the residue examined with the microscope for hæmin crystals; the presence of these

shows the blood belongs to a warm-blooded animal.

3. Another small portion of the red aqueous fluid is placed on a white porcelain dish and a drop of alcoholic solution of guaiacum added, a reddish white precipitate of resin is formed; if on the addition of a drop of peroxide of hydrogen, or autozonized ether, a blue colour is developed, we may suppose that blood is present. But as this blue colour may be developed by guaiacum and ozonizing substances, by casein, gluten, and various iron compounds, the test cannot be relied on unless verified by the other tests.

Sometimes it is necessary to apply the guaiacum and peroxide directly to the stuff that is stained, if so press the moistened surface with white blotting paper, on which the blue stain will appear.

4. A drop of the red solution should be placed on a piece of white porcelain and a drop of dilute ammonia solution added. If the colour does not turn to red or green it shows the stain is not caused by the colouring matter of fruits or flowers.

5. The red colouring matter of blood is destroyed by heat. A small portion of the red aqueous solution should be heated therefore to see if it acquires the brown filmy opacity or coagulation, which denotes the presence of the colouring matter of the blood.

THE CHYLE.

The chyle is the fluid taken up by the lacteals of the intestinal canal, and which circulates through the mesenteric glands, and thoracic duct, and is eventually poured into the general current of the circulation near the junction of the left internal jugular and subclavian veins. In this manner the blood is constantly receiving fresh material derived directly from the intestinal canal.

If examined during the process of digestion, chyle is a white creamy fluid of slight alkaline reaction, rich in oil globules, and like blood separates into clot and serum. The coagulum forming a bulky soft gelatinous clot which after exposure to the air acquires a rosy tint. The serum is turbid, and contains albumin and salts in solution; it is coagulated by heat and acetic acid.

The chyle derived from the intestinal lacteals does not contain fibrin, and consequently does not separate into clot and serum.

If the animal has been fasting, the chyle loses creamy appearance and becomes more transparent and of a yellowish colour.

ANALYSIS OF CHYLE.

	In full digestion.	Fasting.
Water	91·8	96·8
Solids	8·2	3·2
Fibrin	·2	·09
Albumin	3·5	2·30
Fats	3·3	·04
Extractives	·4 ⎫	
Salts	·8 ⎭	·77

The relative proportion between the water and

the solids is subject to variation, the solid matter being in excess during and shortly after digestion.

The albuminous matters consist of albumin casein, and the peptones.

The fatty matters consist of minute spherical globules, and form what Mr. Gulliver calls the *molecular base of the chyle*, their diameter is estimated at $\frac{1}{36000}$ of an inch, they disappear on the addition of ether.

Among the extractives, Urea, Leucin, Tyrosin, and Sugar have been frequently obtained. The ash much resembles that of blood; the sodium having a preponderance over the potassium salts, and the phosphates over the carbonates.

THE LYMPH.

Lymph is a clear, colourless, or straw coloured, transparent fluid of alkaline reaction, and saline taste; it is obtained from the lymphatic vessels and glands. In composition lymph closely resembles chyle, differing chiefly in the smaller proportion of fibrin and fatty matter it contains.

ANALYSIS OF LYMPH.

Water	94
Solids	6
Fibrin	·05
Albumin	4·28
Fats	·38
Extractives . . .	·57
Salts	·72

PUS.

Pus is a pathological fluid, and consists essentially of a liquid portion "liquor puris" which is exuded liquor sanguinis, and white corpuscles or leucocytes, which cannot be distinguished from the white corpuscles of the blood.

ANALYSIS OF LAUDUBLE PUS.

Water	87
Solids	13
Albumins	8·5
Fatty Matters . . .	3·
Extractives	0·7
Inorganic residue . . .	0·8

The proportion of solids to the water, varies of course with the nature of the pus formed; thus, in ichorous, muco, or sero-pus, the solids are diminished.

The Albumins according to Miescher consist of:

(α.) An albumin coagulable at the ordinary temperature of sero-albumin.

(β.) An albumin coagulable at 48-49, the coagula insoluble in hydrochloric acid, in sodium chloride, and also in dilute soda solutions.

(γ.) Alkali albuminate or casein, precipitated by acetic acid.

(δ.) An albumin insoluble in water, soluble in hydrochloric acid, swelling up in solution of sodium chloride; this corresponds to Rovida's hyalin substance.

(ε.) An albumin unaltered in water, sodium chloride solution, and with difficulty soluble in hydrochloric acid $\frac{1}{1000}$.

The extractives contain Urea, Leucin, Cerebrin, Lecithin and Sugar.

The inorganic residue consists chiefly of sodium, potassium, and calcium phosphates, and carbonates with traces of iron, and magnesia. Pus formed in the soft tissues contains often only a trace of calcium phosphate, but pus derived from the neighbourhood of diseased bones often contains 2·5 per cent.

The pus corpuscles can be separated from the liquor puris by the addition of a 10 per cent. solution of sodium chloride, and the precipited mass removed by filtration and thoroughly washed with the same solution till quite free from serum.

The pus-corpuscles are spherical irregular bodies about $\frac{1}{2800}$ to $\frac{1}{3500}$ of an. inch in diameter, containing a number of granules and one or more nuclei. When treated with dilute acetic acid they swell up and become more transparent and the nuclei more distinct. Treated with ammonia or potash solutions, pus becomes tenacious and jelly-like; this character distinguishes it from mucus, which becomes less tenacious and more fluid in the addition of these solutions.

CHAPTER X.

THE MILK.

MILK is the secretion furnished by the mammary gland for the nutrition of the young mammal. It consists essentially of a mixture of three substances, which represent respectively three of the important proximate constituents of the food, viz : 1. Casein, the albuminous; 2. Butter or oily matter, the oleaginous; and 3. Lactose, the saccharine element. Milk also contains a considerable quantity of calcium phosphate, a salt required for the ossification of the bones of the young animal.

Milk is a white or bluish white fluid, having a specific gravity of 1·018 to 1·045; and when fresh an alkaline reaction, due to the large proportion of alkaline phosphates and carbonates it contains, but rapidly becoming acid on exposure to the air from the conversion of *lactose* or milk sugar into lactic acid. After standing some time a yellowish white stratum, the cream, collects on the surface, which is formed by the oily matters coming to the surface and carrying with them a portion of the casein, sugar, and milk salts; the lower portion consequently becomes poorer in quality, and acquiresa more decidedly bluish tint.

Under the microscope milk appears as a turbid

fluid holding a number of fat globules in suspension; these globules are about $\frac{1}{1200}$ to $\frac{1}{1800}$ of an inch in diameter, and are not dissolved, unless agitated, by alkalis or acetic acid, neither are they dissolved when agitated with ether unless caustic potash is added; from this fact it has been supposed that the globules are surrounded by a caseous or albuminous envelope.

Milk is not coagulated by heat at a temperature of 100°, though a pellicle consisting of a combination of casein with the inorganic salts forms on the surface. Exposed for some time at a temperature of 40° C., milk undergoes alcoholic fermentation, the casein converting one part of lactose into lactic acid, and the remainder into grape sugar which is in turn converted into alcohol. It is by this means the Tartar tribes obtain their *koumis*. Milk is coagulated by mineral acids, by acetic and other organic acids, and by rennet a substance obtained from the fourth stomach of ruminant animals; the quantity of these substances necessary to produce immediate coagulation depends on the amount casein present.

AVERAGE COMPOSITION OF MILK.

	Human.	Cows.
Water	881	870
Casein	29 .	40
Butter	35	31
Sugar	50	54
Extractives	2	2
Salts	3	3

WATER. Evaporate 5 grms. of milk, in a thin weighed platinum dish, in a water bath for about 3 hours at a temperature of 100° C. till the weight of the residue becomes constant. The loss of weight during the evaporation will give the amount of water. A fair sample of milk should yield a residue of total solids of 12-13 per cent.

SALTS. The dry residue, after the evaporation of the water, is burnt over a Bunsen's lamp and incinerated till the organic matter is burnt off and the ash becomes nearly white. The weight of the ash represents the quantity of inorganic matter. The ash yields about 70 per cent. of calcium, magnesium, potassium, and sodium phosphates; 7 per cent. of alkaline carbonates; 14 per cent. of potassium chloride; 5 per cent. of sodium chloride; and 4 per cent. of other salts in which are found traces of calcium fluoride, potassium silicate and iron phosphate.

CASEIN FATTY MATTERS AND LACTOSE. 50 grms. of milk is boiled for some time with a little dilute acetic acid and the coagulum collected on a filter and thoroughly exhausted with ether, and the etherial washings allowed to fall into the filtrate. The coagulum is then mixed with one fifth its weight of powdered calcium sulphate, evaporated to dryness in a water bath, and the residue powdered and exhausted with ether. The residue is again dried and weighed, the weight representing the amount of casein in 50 grms. of milk, minus the weight of calcium sulphate. Wanklyn suggests the "ammonia-process" as the best method of determining the casein in milk, 6·5 parts of ammonia being furnished by the oxidation of 100 parts of casein when boiled with alkaline permanganate. (*Milk Journal*, vol. i. p. 160.)

To the filtrate, obtained by the removal of the coagulum, and into which the etherial washings have been returned, add some more ether and frequently agitate, pour off the etherial solution and evaporate, the residue will represent the fatty matter. Marchand recommends the following ready method of estimating the fatty matters in milk. He divides a test tube into three divisions, marked A. B. c. respectively, each capable of holding 10 C.C.'s.. The portion of the test tube C. *a.* is filled with pure milk to which has been added two drops of solution of soda; the portions A. B. and B. c. then filled with a mixture of alcohol and ether and the whole mixture in the tube well agitated and then the tube is plunged into a water bath at a temperature of 40° C., and after some time the fatty matters separate and collect at the upper part of the tube A. B. which is divided into 50ths, and according to the richness of the milk, so will be the yield of fatty matters in the upper portion of the tube.

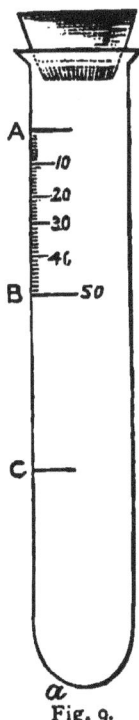
Fig. 9.

The *cream*, which consists of a mixture of the oily matters with a portion of the casein, sugar, and milk salts, is estimated by filling a glass cylinder graduated to 100 parts with new milk up to 0°, after some hours the cream rises, and the per centage amount can then be read off. The quantity of cream in good milk should never fall below 5 per cent. .

To ESTIMATE THE SUGAR. Fehling's alkaline copper solution must be applied to the liquid left after the withdrawal of the casein and fatty matters, in the manner directed in making the quantitative estimation of sugar in urine.

THE COLOSTRUM. The milk secreted the first week after delivery has a yellowish tint, and is richer in casein, fatty matter, and sugar than that passed at a subsequent period. Examined under the microscope, in addition to the ordinary milk globules, a number of granular corpuscles are observed, called colostrum corpuscles; they are formed by the fatty degeneration of epithelial cells lining the mammary ducts. Colostrum also contains a small quantity of albumin coagulable by heat. Specific gravity 1·045 to 1·050 and of strongly alkaline reaction.

ANALYSIS OF COLOSTRUM AND ORDINARY HUMAN MILK.

	Colostrum.	Ordinary Milk.
Water	835	881
Casein	40	29
Butter	50	35
Sugar	70	50
Salts and Extractives	5	5

SEMINAL FLUID.

This fluid is secreted by the testicles, before emission it is mixed with the secretion of the prostate and Cowper's glands, it is therefore difficult to obtain it unmixed for chemical examination. A very good idea can be obtained however of its composition, by carefully picking out the tubuli seminiferi of the testicle of an animal removed in the rutting season, and squeezing out the contents.

ANALYSIS OF THE EXPRESSED JUICE OF THE TUBULI SEMINIFERI.

Water . . .	86.
Solids . . .	14
Albumins . . .	6·5
Extractives . .	1·5
Fats	2·5
Salts	3·5

THE ALBUMINOUS MATTERS consist of ordinary albumin which is deposited from an aqueous solution by a temperature of 73°. An albumin precipitated from an aqueous solution by acetic acid, and which therefore resembles alkali albumin or casein. Globulin which is deposited from a sodium chloride solution of the gland by carbonic acid gas. And myosin which is thrown down from the same solution by the addition of a dilute acid. Mucin is also obtained in small quantities by the addition of acetic acid to an alkaline solution. The peculiar albuminoid substance, *spermatin*, is probably a mixture of globulin and lecithin.

THE EXTRACTIVES obtained from the aqueous and alcoholic extracts, are leucin, tyrosin, urea, kreatin, and inosite. Dr. Treskin has found a peculiar organic acid the composition of which he does not give; the same observer has also found kreatinin, probably formed by the decomposition of kreatin.

THE FATS consist principally of a mixture of stearin with palmitin, and olein, together with lecithin, cerebrin, and cholesterin.

THE SALTS consist principally of sodium and potassium chlorides, potassium sulphate, and especially of calcium and magnesium phosphate.

FLUIDS OF THE MAMMALIAN OVUM.

The liquor amnii and allantoic fluid are usually clear and without colour. The proportion of solids to the water is extremely small. They contain traces of albumin, about 0·1 to 0·15 per cent. A considerable quantity of sugar; and a variable amount of urea. The allantoic fluid contains a characteristic ingredient, *allantoin*.

The inorganic constituents of both fluids consist chiefly of sodium chloride, and calcium and magnesium phosphate.

PART V.

THE EXCRETA.

THE Excreta consist of the ultimate products of the decomposition of the Saccharine, Oleaginous, and Albuminous principles, which having played their part in providing for the nutrition of the body, and being no longer of service, are cast out.

These products; viz. Water, Carbonic Acid, and certain Nitrogenous bases, which we have considered individually in Chapter V., are eliminated by four channels; viz., 1. The Lungs. 2. The Skin. 3. The Bowels. 4. The Kidneys; in about the following proportions, calculated in grammes.

	Lungs.	Skin.	Bowels.	Kidneys.
Water	311·	660·	128·	1766·
Carbonic Acid .	953·	6·		
Nitrogen .	·04	0·7	3·	15·

Besides the products of decomposition, the cutaneous, fæcal, and urinary excreta contain the inorganic substances which were introduced with the food; thus, by the sweat about 1·2 grm. of fixed salts are eliminated in the 24 hours, by the fæces 26· grms., and by the urine 6 grms.

CHAPTER XI.

THE BREATH.

RESPIRATION is the process by which the venous blood becomes aerated by the absorption of oxygen, and the liberation of carbonic acid. In man this process is effected by the lungs, and to a slight extent by the skin.

The nature of the interchange between the two gases is partly chemical, and partly mechanical; thus, the absorption of oxygen is effected by its chemical union with the hæmoglobin of the blood, forming oxy-hæmoglobin, the formula of which, according to Dr. Preyer, is $\left.{\begin{matrix}O\\O\end{matrix}}\right\}$ Hb. The escape of the carbonic acid is produced by both means, since carbonic acid is present in the blood in two conditions; viz., loose and stable. By loose carbonic acid is meant that carbonic acid which is given off to a vacuum, and is physically absorbed, as well as retained, in solution by the alkaline phosphates and carbonates: this loose carbonic acid escapes from the lungs into the atmosphere in obedience to the law of pressures. On the other hand the stable carbonic acid is in combination with some base, and can only be separated from the blood by chemical means. How the decomposition is effected we cannot yet determine, but Dr. Preyer has suggested that it is effected in some way by the union of oxygen with hæmoglobin, by which the fixed $_4CO_2$ is converted into loose CO_2, by the probable development of some acid.

The quantity of air passing through the lungs in the 24 hours is subject to great variation, depending on the habits, health, and general conditions of the individual. Dr. Ed. Smith found that an unoccupied gentleman respired 804,780 cubic inches, an ordinary tradesman 1,065,840, and a hard-worked labourer 1,568,390 cubic inches.

The expired air, or breath, differs from the inspired by having a higher temperature, by containing more aqueous vapour, by the increase of the carbonic acid, and the proportionate decrease of the oxygen; it contains also slightly more nitrogen and also traces of marsh gas, and some volatile matters. The expired air also occupies a greater volume than the inspired. This is owing to the higher temperature, and the presence of aqueous vapour. For if it be dried, and the temperature reduced to the level of that of the external air, a diminution of volume will be noticed, corresponding to the amount of oxygen (about one-seventh of the whole quantity) which is not converted into carbonic acid, but is required for the oxidation of hydrogen, sulphur, etc., within the body.

THE TEMPERATURE of the expired air in health is about 35—36° C., this is considerably higher than the usual temperature of the atmosphere.

THE AQUEOUS VAPOUR. The quantity of aqueous vapour given off by the lungs in the 24 hours depends greatly upon the temperature, the humidity, and pressure of the atmosphere: the average, however, may be taken as 311 grammes. It is estimated by passing the expired air through bent tubes, containing sulphuric acid and pumice stone, the weight of the whole apparatus being ascertained. The increase of weight of the apparatus, after the breath has been passed through it for some time, denotes the amount of aqueous vapour which has been absorbed by the sulphuric acid.

CARBONIC ACID. According to Vierdot about 4·7 per cent. of the Oxygen is removed from the air, in exchange for 4·3 per cent. of Carbonic acid, at the ordinary rate of respiration.

Dr. Ed. Smith taking 1,000,000 cubic inches to represent the average total amount of respired air in the twenty-four hours, calculates the quantity of Oxygen withdrawn from it at 1223 grammes, and the Carbonic acid exhaled at 953 grammes.

The amount of Carbonic acid exhaled in the twenty-four hours varies greatly being influenced by age, sex, and health of the individual, and also by certain external influences, as the temperature of the outer air, the season of the year, the period of the day, the nature of the food, and the amount of bodily exercise taken. These variations have been thoroughly investigated by Dr. Ed. Smith, and the result of his work may be thus briefly stated.

1. Age. The amount of Carbonic acid exhaled is greatest in adults between 30 and 45 ; it is least in youth and old age.

2. Sex. Females exhale less Carbonic acid, pro-

portionately, than males; there is also a decrease during the menstrual periods; there is a slight increase after the final cessation of menstruation, but the quantity is again diminished as the female gets older; during pregnancy there is an increase in the exhalation.

3. Health. In certain diseases, as the exanthe mata, chronic pulmonary and skin diseases, chlorosis, etc., the quantity is increased.

4. Temperature. The elimination of Carbonic acid is increased by cold and diminished by heat.

5. Season. The elimination is greater in the spring and autumn months than in the winter.

6. Period of the day. The maximum elimination is before and after noon, the minimum before and after midnight. The quantity is increased after meals.

7. The nature of the food. All nitrogenous food as casein, albumin, gelatin, also tea, coffee, sugar, rice, oatmeal and alcohol increased the elimination.

8. Muscular exertion. The elimination of Carbonic acid is greatly increased by exercise. Dr. Ed. Smith found the quantity evolved whilst walking at 3 miles an hour was 1·10 to 1·67 grms. per minute whilst with increased exertion it rose to 2·7 grms.

Estimation of the Carbonic Acid in the expired air. Dr. Ed. Smith in his experiments used an apparatus, which may be thus briefly described. The air for inspiration is conducted by means of a caoutchouc tube from a small spirometer, capable of registering from one to a million cubic inches to a mask which closely covers the nose and chin. The expired air is prevented from passing back into the spirometer by means of a valve, but is directed by another tube into the analytical apparatus. This consists: first, of a desiccator of sulphuric acid and pumice stone, of a capacity of 70 cubic inches, in which the expired air is dried before passing into the second part of the apparatus, or potass box. This is made of gutta percha with dimensions of 12 in. x 12 in. x 5 in. and is divided into 5 chambers and each chamber is sub-divided into six cells by means of strips of gutta percha; the chambers communicate with each other, and into each of which the fluid potass is passed by means of gutta percha tubes. Thirty fluid ounces of solution of caustic potass (sp. gr. 1·27) are introduced into the box, and this quantity of potass absorbs 600 grains or 38 grms. of Carbonic acid. The air after passing through the potass chamber, enters another dessicator similar to the first where the watery vapour carried off from the fluid potass is absorbed. The increase in weight of the mask, expiratory tube, and first dessicator, gives the amount of aqueous vapour in a known volume of expired air. And the increase in weight of the second part of the apparatus, namely the potass box and second dessicator, gives the weight of the Carbonic acid.

NITROGEN. The increase in the amount of Nitrogen in the expired air is extremely small, not amounting to more than 0·40 grm. in the 24 hours; this is probably derived from the decomposition of animal matter lodged in the teeth and air-passages.

VOLATILE MATTERS. The expired air contains small quantities of ammonia, traces of marsh gas probably derived from the intestines, and more or less animal matter in a state of decomposition. In disease, sodium chloride, uric acid, and ammonium urates have been discovered in the breath.

SWEAT.

Aqueous vapour is constantly being carried off by evaporation from the surface of the body, and therefore under ordinary circumstances is not deposited in the liquid form; this constant and imperceptible removal of the aqueous vapour is known as the *insensible transpiration*. When however the quantity of the secretion is increased, fluid collects in drops on the skin, this fluid or sweat is known as the *sensible transpiration* of aqueous vapour.

The sweat is a colourless fluid, of acid reaction and a specific gravity of 1·004; containing epithelial scales, sebaceous matter, and volatile fatty organic acids.

The average quantity of aqueous vapour discharged by the skin in the 24 hours is about 660 grammes or about 2 lbs. This quantity is subject to considerable variations; it is increased by heat and a dry atmosphere; also by the ingestion of food, and by moderate exercise; it is diminished by cold, an atmosphere charged with aqueous vapour, by fatigue, and exhaustion.

According to Funke about 0·7 grm of nitrogen, is daily eliminated by the skin in the form of urea.

The average exhalation of carbonic acid from the skin in the 24 hours amounts to about 6 grammes.

COMPOSITION OF SWEAT.

Water	995
Solids	5
Epithelium	1·2
Organic Acids	·9
Fat	·7
Extractives.	·4
Salts	1·8

THE VOLATILE ORGANIC ACIDS, consist of formic, acetic, and butyric acid and sometimes caproic and capric acids, are separated and estimated by fractional distillation, as described in Chapter IV.

THE FATTY MATTERS, contain a large proportion of stearin derived from the secretion of the sebaceous glands. They are estimated by exhausting the residue left after fractional distillation with ether and evaporating the etherial solution in a weighed platnum capsule and ascertaining the weight of the residue.

THE EXTRACTIVES. The aqueous and alcoholic extract obtained from the residue left after fractional distillation, always yields a small quantity of urea;

in disease this quantity is increased, aud sometimes uric acid, leucin and sugar are met with.

SALTS.   Sodium Chloride constitutes nearly one half the ash ; the earthy phosphates and a trace of iron oxide which are present, are derived from the epithelium.

THE FÆCES.

The fæces consist of the undigested and insoluble residue of the food, mixed with some of the products of the biliary and intestinal secretions.

The quantity passed daily by a healthy adult depends much on the amount and nature of the food ingested, the average, however, may be stated at from 7 to 8 ounces.

According to Dr. Ed. Smith, 3· grms. represents the average daily elimination of nitrogen by the bowels.

The colour of normal fæces is dark brown, due chiefly to the presence of the colouring matter of the bile, and a red colouring matter derived probably from decomposed hæmatin. When little or no bile is passed into the intestines the fæces acquire a pale ashy colour. Certain articles of diet or medicine impart different colours to the fæces ; thus, iron makes them black, and mercury a deep olive green.

The peculiar odour of fæces is not derived from the decomposition of the undigested residue of the food, but seems to be due to a peculiar animal matter possessing a putrescent odour and which is

eliminated by the glandulæ of the intestinal canal.
Meconium, or the fæcal looking matter, found in
the intestinal canal at birth consists almost entirely
of biliary matters.

ANALYSIS OF MECONIUM.

Biliary resin   15·6

Cholesterin ⎫
Olein   ⎬   15·4
Palmitin   ⎭

Epithelium ⎫
Pigment   ⎬   69·
Salts   ⎭

COMPOSITION OF FÆCES.

Water	.	.	.	.	·73
Solids	.	.	.	.	·27
Albumin	.	.	.	.	0·9
Mucus	.	.	.	.	4·0
Extractives	.	.	.	.	2·8
Fat .	.	.	.	.	1·5
Salts	.	.	.	.	1·8
Resinous biliary and colouring matters		.	.	9·0
Insoluble residue of food			.	7·0

WATER. A small portion of fæces is evaporated
in a weighed platinum dish over a water bath, and
the loss of weight gives the amount of water driven
off; this generally is from 73 to 75 per cent.

SOLIDS. The dry residue is incinerated, and the
weight of ash gives the quantity of salts, whilst this

weight deducted from the weight of the total residue, gives the weight of the organic solids.

ALBUMIN. A small quantity of coagulable albumin is always present in normal fæces, and in cases of dysentery, typhus, and cholera the quantity passed with the evacuations is considerably increased.

EXTRACTIVES. Besides the occasional presence of urea, uric acid, leucin, etc., three substances, stercorin (p. 21), excretin (p. 23), excretolic acid (p. 23) can always be obtained from fæces and seem to be peculiar to it.

FATS. The fæces are exhausted by boiling alcohol, and the alcoholic solution filtered, and allowed to cool, when a deposit forms which is treated with boiling alcohol which partially dissolves it. The insoluble portion of the deposit is then boiled with potash and the alkaline solution neutralized with hydrochloric acid, when oleic, stearic, and palmitic acids will be obtained.

SALTS. The residue left after incineration according to Enderlin has the following composition.

ANALYSIS OF 100 PARTS OF FÆCAL ASH.

Alkaline Chlorides and Sulphates .	1·7
Neutral Sodium Phosphate . .	2·8
Alkaline Calcium and Magnesium } Phosphate . . . . }	80·5
Iron Phosphate	2·4
Calcium Sulphate	4·7
Silica . . . . . ' .	7·9 ·

According to Berzelius more lime than magnesia

is absorbed in the intestine, since in the fæcal ash less lime and relatively more magnesia is found than in the food which has been ingested: the ratio of magnesia being as 1 : 2 or 1 : 2 ·5 in the fæces·

The potassium salts are greatly in excess of the sodium, and this excess is increased by an abundant meat diet.

BILIARY MATTERS appear in the fæces in an altered condition, the bile pigments and acids being de-composed, the latter furnishing dyslysin, cholalic acid, and choloidinic acid.

THE INSOLUBLE RESIDUE OF THE FOOD consists chiefly of cells of cartilage, fibro-cartilage, fibres of elastic tissue, undigested muscular fibres, the outer envelope of vegetable cells and fibres, and partially dissolved starch granules; mixed with this undi-gested mass crystals of ammonia magnesian phos-phate are frequently to be recognized under the microscope.

THE URINE.

Healthy human urine is a clear, transparent, amber-coloured fluid, of faint aromatic odour, and of a saltish bitter taste.

Its reaction is acid, which is due to the presence of the acid sodium phosphate, free lactic acid, and some other organic acids. The intensity of this reaction increases in urine that has been passed some hours, owing to the lactic and butyric acid fermentation of the extractive matters. When decomposition of the urea begins, the reaction becomes alkaline from the formation of ammonium carbonate. The urine passed immediately after a meal is less acid than that passed before taking food, indeed the reaction is then often neutral or even alkaline.

The specific gravity of the twenty-four hours urine may be said to vary from 1·016 to 1·024. The specific gravity of urine passed soon after eating is high; viz., 1·020 to 1·030; on the other hand after the ingestion of much fluid the specific gravity may sink as low as 1·002. The specific gravity of the "urina sanguinis," or urine passed in the morning, is always high. In the latter months of pregnancy the specific gravity is considerably increased.

The mean average quantity of urine passed by a healthy adult may be stated at 52½ fluid ounces, but the secretion may fall as low as 30 ounces or rise to 80 ounces and still be within the limit of health. The various conditions that influence the quantity of urine passed in health, in the twenty-four hours may be thus briefly stated.

1. More urine is passed in cold than in hot weather.

2. The ingestion of fluids, or any of the urinary constituents as sodium chloride, urea, etc., increase the amount of urine. So also does the ingestion of highly animal food, from its containing one of the principle elements of urinary excretion; viz., nitrogen.

3. Children pass more urine proportionately than adults; and men more than women; in old age the secretion is diminished.

4. Exercise by increasing the pulmonary and cutaneous exhalation, diminishes the water of the urine. The urine is the principal channel provided for the elimination of nitrogen, about 15 grms. of which, as an average quantity, are passed into the urine in the 24 hours, in the form of urea, uric acid, kreatinin, etc. The quantity of nitrogen passed into the urine is increased by an animal diet.

COMPOSITION OF HEALTHY HUMAN URINE.

Water	940
Solids	60
Urea	31·
Uric acid	0·45
Kreatin	1·20
Kreatinin	1·50
Xanthin	traces
Hippuric acid	0·25
Organic acids*	0·12
Pigment and Mucus . .	0·35
Acid sodium phosphate	
Neutral sodium phosphate	
,, potassium ,,	9·50
Calcium phosphate	
Magnesium ,,	
Potassium chloride	
Sodium ,,	8·50
Potassium sulphate	7·0
Iron. Silica. Fluorin.	traces

* These acids consist of Phenylic, Damaluric, Damolic, Taurylic, and Kryptophanic acids; also a fatty acid which is probably palmitic.

THE WATER AND SOLIDS are determined by placing 20 C.C. of urine in a small platinum dish of known weight, and ascertaining the weight of this quantity of urine. The dish is then placed over the water bath, and the contents evaporated till it ceases to lose weight. The weight is again taken, and the loss of weight represents the amount of water which has been driven off; and the weight of the residue, the amount of total solids. By the incineration of this residue the proportion of the inorganic salts is determined.

ESTIMATION OF UREA. The urine is collected for 24 hours and carefully measured;* of this urine measure off with a pipette 20 C.C. and precipitate the phosphates and sulphates by the addition of 20 C.C. of a saturated solution of barium nitrate and hydrate. Filter, and take of the clear filtered solution 20 C.C., which contains 10 C.C. of urine, and add to it a few drops of solution of silver nitrate to precipitate the chlorides, and allow the precipitate to subside to the bottom of the vessel containing the urine. Now fill a Mohr's burette with 50 C.C. of the standard solution of "mercuric nitrate,"† and allow it to fall gradually into the urine. Add first of all 5 C.C. of mercuric solution, stirring well with a glass rod, then remove a drop of the mixture and let it fall on a piece of paper saturated with sodium carbonate; if no yellow stain is given, add 5 C.C. more of the

* If albumin is present in the urine, it must be coagulated and removed by filtration.

† See Appendix.

mercuric solution, and again test; if there is no result, add 1 C.C. of the solution at a time till the yellow colour appears on the test paper; when this is the case the process is completed, and the estimation can be made. Thus, since each C.C. of the mercuric solution precipitates ·01 grms. of urea, therefore the number of C.C.'s of the mercuric solution used will denote the quantity of urea in 10 C.C. of urine; and if the total quantity of urine passed in 24 hours be multiplied by the number of C.C.'s of mercuric solution used, and divided by 10, the quantity of urine submitted to analysis, the amount of urea eliminated in the 24 hours will be obtained; for example, 1520 C.C. of urine are passed in 24 hours, 10 C.C. are used for analysis, and 24 C.C. of mercuric solution are employed to precipitate it; then, $\dfrac{1520 \times \cdot 24}{10} =$ 36·48 grms. of urea.

ESTIMATION OF URIC ACID. Collect the urine for 24 hours, measure carefully; of this urine take 200 C.C., place it in a tall urine glass, and add 20 C.C. of strong hydrochloric acid. The mixture is then set aside in the cool for 24 hours, at the end of this time crystals of uric acid will be deposited on the sides and bottom of the glass. Collect these crystals on a small weighed filter, and wash with a few drops of alcohol acidulated with hydrochloric acid; remove the filter with the crystals to the air bath, dry, and then weigh. Then the weight of the crystals thus obtained, multiplied by the whole of the 24 hours urine, and divided by

200 C.C. the quantity of that urine employed, will give the amount of uric acid eliminated in the 24 hours. Thus, if the quantity of urine passed in the 24 hours be 1000 C.C., and the amount of uric acid obtained from 200 C.C. of this urine be ·08 grm.; then ;

$$\frac{1000 \times ·08}{200} = ·4 \, grm.$$

The uric acid thus obtained is not pure, being inseparably mixed with a variable quantity of colouring matter, which of course increases the weight; this however is counterbalanced by the fact that a small quantity of uric acid always remains unseparated in the urine.

Refer, for separation of Kreatin, Kreatinin, Xanthin, Hippuric Acid, and the Organic acids, to chapters IV. and V.

THE PIGMENT. The yellow colour of the urine depends upon the presence of a pigmentary extractive, the exact nature of which has been the subject of much discussion.

According to Dr. G. Harley, this colouring matter closely resembles the hæmatin of the blood, and he gives it the name of uro-hæmatin; he directs it to be prepared as follows. Evaporate a large quantity of urine to the consistence of treacle, removing the sodium chloride, and other salts as they crystallize. The colouring matter is extracted from this molasses like syrup by alcohol. The deeply-tinged alcoholic extract is now boiled and treated with slaked lime until it is completely decolorized; it is then filtered, and the compound of lime and colouring

O

matter well washed with water, and afterwards with ether, to free it from fat, of which there is invariably a considerable quantity. The lime compound, when dry, is decomposed by hydrochloric acid and extracted by boiling alcohol. The alcoholic solution is now mixed with an equal portion of ether and after being frequently shaken, allowed to stand a day or two, in order that the ether may dissolve as much as possible of the pigment. On the addition of water, the ether charged with colouring matter separates, and is decanted. This etherial solution is treated with chloroform, which on evaporation deposits pure uro-hæmatin. This substance is a red, non-crystallizable, organic compound, somewhat resembling soft red sealing-wax; it is insoluble in water and in solution of sodium chloride, but soluble in chloroform, ether, alcohol, and urine; when burned the ash yields an abundance of iron oxide.

Dr. Ed. Schunk considers that human urine contains under all circumstances two distinct colouring extractive matters.

(*a.*) Soluble in alcohol and ether.

(*b.*) Soluble in alcohol but not in ether.

To the substance, soluble in alcohol and ether, he gave the name urian, and assigned to it the formula of $C_{43}H_{51}NO_{26}$; and to the substance insoluble in ether, he gave the name urianin and the formula $C_{19}H_{27}NO_{14}$. The relation in which urian stands to urianin may be represented by this formula.

Urian.   Urianin.   Glucose.

$$C_{43}H_{51}NO_{26} + 12H_2O = C_{19}H_{27}NO_{14} + 4(C_6H_{12}O_6).$$

The oxidation of these substances gives rise to the blue (uroglaucin), the red (urohodin), and the black (uromelanin) colouration of the urine often observed in diseased states of the system. (See Mr. Schunk's paper in *Proceedings of Royal Society,* 1867).

Dr. Thudichum gives the name of urochrome to the yellow colouring matter of the urine. This substance, he says, is soluble in ether, less so in alcohol, and under various processes of decomposition yields a red resin uropithin, also uromelanin, and numerous other products. He considers urochome to be derived from albuminous matter, but to have no immediate relation with the colouring matter of the bile or blood.

Notwithstanding the difference of opinion, regarding the exact nature of the urinary colouring matter, its general characters may be thus broadly stated.

1. It is an albuminous substance, containing a small proportion of nitrogen, and associated with a certain amount of iron which is probably combined with it.

2 The albuminous matter strongly resembles Indican, for like that substance it undergoes a change of colour, to blue, green, red, under different conditions of oxidation.

3. Most pathologists now consider that this, Indican-resembling substance, is derived from the destruction of the hæmoglobin the well known colouring matter of the blood. For an increase of the pigmentary matters in the urine can always

be demonstrated in cases of anæmia, chlorosis, albuminuria, etc.; diseases associated with the destruction of a large number of blood corpuscles.

ESTIMATION OF THE ALKALINE AND EARTHY PHOSPHATES. Collect the urine passed during the 24 hours, and carefully measure; of this urine measure off by means of a pipette, 50 C.C. into a small beaker, and add 50 C.C. of saturated sodium super acetate solution and heat the mixture in a water bath to 90° or 100° C. Then add from a Mohr's burette, the standard solution of Uranium nitrate* till a precipitate is no longer formed, and a potassium ferrocyanide test paper is stained brown when touched by a drop of the mixture. As 1 C.C. of the uranic oxide solution corresponds to ·005 grm. of phosphoric acid; therefore if 17 C.C. if uranium nitrate solution be required to precipitate the 50 C.C. of urine employed, then $17 \times ·005 = ·085$ grm. of phosphoric acid; and if the patient passed 1250 C.C. of urine in the 24 hours, then that quantity multiplied by ·085 grm. the amount of phosphoric acid precipitated from 50 C.C. of urine, and divided by 50 C.C. the quantity of urine used for analysis, will give $\dfrac{1250 \quad ·085}{50} = 2·12$ grms. of phosphoric acid in the 24 hours urine.

To estimate the Earthy Phosphates separately, they must be removed from the Alkaline Phosphates by precipitation with ammonia, collected on a filter and well washed with dilute ammonia. The filter is then broken, and the precipitate dissolved in as

* See Appendix.

small a quantity of acetic acid as possible and distilled water added to 50 C.C. ; complete the estimation as directed above for the total phosphates.

Separate estimation of Lime and Magnesia see Part III.

ESTIMATION OF THE CHLORIDES. Collect the urine for 24 hours, carefully measure ; of this urine measure off, by means of a pipette, 50 C.C. into a small beaker, and add a few drops of sodium carbonate solution, to render it neutral. A few drops of potassium chromate solution are now added and a few C.C.'s of the standard solution of silver nitrate° run into the mixture from a Mohr's burette, and agitate ; continue to add a C.C. or so of the standard solution till a red colour appears when the mixture is agitated (red silver chromate). Now since 1 C.C. of the silver nitrate solution is equal to ·006 grm. of hydrochloric acid, therefore if 17 C.C. of silver nitrate be used, the 50 C.C. of urine will contain ·102 grm. of hydrochloric acid and if 1500 C.C. of urine be passed in the 24 hours, then

$$\frac{·102 \times 1500}{50} = 3 ·02 \text{ grms. of hydrochloric acid.}$$

ESTIMATION OF SULPHATES. Collect the urine for 24 hours, and carefully measure ; of this urine measure off, by means of a pipette, 50 C.C. into a beaker, add 10 drops of hydrochloric acid, and boil. Then from a Mohr's burette allow the standard solution of barium chloride° to run into the mixture till a precipitate ceases to be formed. Now as 1 C.C. of the barium chloride solution

* See Appendix.

equals ·01 of sulphuric acid therefore if 10 C.C.'s
of this solution are required the quantity of sulphuric acid in 50 C.C. of urine will be 0·1 grm.;
and by multiplying this quantity by the total
amount of the 24 hours urine, say 1200 C.C., and
dividing by 50 C.C. the quantity of urine employed

for analysis; we have, $\dfrac{0\cdot1 \times 1200}{50}$ = 2·4 grms. of

sulphuric acid passed in the 24 hours.

ESTIMATION OF FREE ACIDS.   Place in a mixing
jar 100 C.C. of urine and add, from a fine burette
graduated to show the 10ths of C.C., the standard
soda solution,° drop by drop.   After each addition shake the mixture and test its neutrality with
litmus paper.   When the mixture is quite neutral
no more Soda Solution is to be added.   Now as
every C.C. of the Soda Solution used to effect this
neutralization corresponds to ·01 grm. of crystallized oxalic acid.   Therefore, if 3 C.C. of Soda
Solution be used to neutralize the urine the
degree of acidity of that fluid will be ·03.

MORBID URINE.

Any of the preceding normal constituents of the
urine may be increased or diminished by abnormal
conditions of the system, and in addition there
may be one or more ingredients present which are
not met with in the healthy secretion.

I. ALBUMIN.   This substance when present in
urine, is detected by adding a few drops of acetic
acid, carefully avoiding excess, to that fluid and

* See Appendix.

then boiling; when the albumin is deposited in floculent coagula, which are insoluble in nitric acid.

If heat alone were employed, the deposition of earthy phosphates may be mistaken for albumin, these however remain in solution when acid is added. When the urine is itself strongly acid there is no need of adding acetic acid. When the urine is alkaline, acetic acid should be added till the urine acquires a decidedly acid reaction.

Sometimes if cystin be present the acetic acid precipitates it, the solubility of this deposit in nitric acid distinguishes it from coagulated albumin.

Estimation of Albumin in Urine. Collect the urine for twenty-four hours and measure; introduce 50 C.C. of this urine into a Mohr's burette and allow it to fall, a C.C. at a time, into a porcelain dish containing an ounce of boiling distilled water. If the urine is sufficiently acid of itself no further addition of acid will be required, but if not it will be necessary when all the urine has been passed into the boiling water, to add a drop or two to the mixture, most carefully avoiding excess. When the albumin is completely coagulated, it is allowed to settle at the bottom of the vessel before proceeding to filtration. When the supernatant fluid is quite clear, it is poured upon a weighed filter, the coagulated albumin remaining on the filter whilst the fluid runs through; any particles of albumin adhering to the porcelain dish are to be removed with a feather and placed on the filter. The mass is then well washed with boiling water till the washings give no precipitate with silver nitrate. The filter with the mass is now removed and placed in a watch glass and carefully evaporated over a water bath until it ceases to lose weight. The whole is then carefully weighed, and after deducting the original weight of the filter and watch glass from the total weight, the remainder represents the quantity of albumin in 50 C.C. of urine, and if this quantity be 0·3 grms. and the amount of the twenty-four hours 1200 C.C., then $\dfrac{1200 \times \cdot 3}{50} = 7\cdot2$ grms. of albumin in the twenty-four hours urine.

2. SUGAR. Diabetic urine is usually of high specific gravity, and of a paler colour than healthy urine.

Boiled with Liquor Potassæ it acquires a deep brown tint. (Moore's Test).

Heated with a few drops of Liquor Potassæ and dilute Cupric Sulphate solution, it reduces the latter and deposits it as a red precipitate of cuprous oxide. (Trommer's Test.)

Boiled with an equal quantity of Sodium Carbonate solution to which a small quantity of bismuth nitrate has been added, the mixture becomes dark coloured from reduction of the bismuth; this sinks to the bottom and forms a greyish black precipitate : if the urine does not contain sugar the precipitate is white. (Bötteger's Test.)

A little yeast added to diabetic urine immediately sets up vinous fermentation, with the liberation of carbonic acid gas.

The surface of diabetic urine if exposed for a short time to the air becomes covered with a scum, which under the microscope appears to consist of minute oval vesicles which are the sporules of Torula Cervisæ, these sporules are commonly joined together so as to form an irregularly jointed confervoid stem.

By evaporating diabetic urine, flat elongated crystals, frequently arranged in stellate or arborescent tufts are obtained, these crystals consist of diabetic sugar with sodium chloride.

Estimation of Sugar in diabetic urine. (Fehling's Method.) The urine for examination must be collected for a period of twenty-four hours, and carefully measured. Filter, and dilute 10 C.C. of the measured urine with distilled water to the volume of 200 C.C., and with this mixture fill a Mohr's burette. Now take a porcelain evaporating dish capable of holding 3 ounces, and place in it, first 30 C.C. of distilled water, then 10 C.C. of the standard Cupric Sulphate Solution* and lastly 10 C.C. of the standard* Alkaline Solution.

This mixture is to be gradually raised to the boiling point, and during the whole process it must be kept at a temperature just short of ebullition. A few C.C.'s from the burette should now be allowed to fall into the warm alkaline cupric solution, which at once becomes turbid; as more urine is added the colour becomes redder, and a

* See Appendix.

precipitate begins to settle readily at the bottom of the capsule. At this point greater care must be exercised in adding the urine, the addition of 1 C.C. at a time being sufficient ; after each addition the capsule should be slightly tilted to observe better if the blue tint of the liquid has disappeared. When the liquid becomes quite colourless the operation is complete, and the estimation can be made. This is done by dividing the whole quantity of urine passed in the twenty-four hours, by the number of C.C.'s of dilute urine from the burette required to effect the complete reduction of the copper solution. For example, a patient passes 4110 C.C. of urine in the twenty-four hours and 30 C.C. of dilute urine are required to reduce the copper solution, then the quantity of sugar will be $\frac{4110}{30} = 137$ grms. in the 24 hours.

Since uric acid also has the power of reducing cupric sulphate, it must be removed if present in any quantity, or it will interfere with the accuracy of the result. Fehling recommends this to be done by precipitating it with lead acetate ; if the urine be diluted before the lead acetate is added, no precipitation of sugar will occur.

Albumin prevents the separation of the cuprous oxide, it should therefore be removed by coagulation if present in the urine.

Roberts' Fermentation Test. The urine is collected for 24 hours and carefully measured, and 4 ounces of this taken and placed in an 8-ounce bottle together with a small piece of yeast, and in another bottle a similar quantity of urine but no yeast. The two bottles are now to be put aside in a warm place for 24 hours and the contents of each having been poured into two urine glasses their respective specific gravities are to be taken. *The difference of each degree lost, in the urine which has the yeast, indicates the presence of one grain of sugar in every fluid ounce of urine.* For example, a patient passes 160 ounces of urine in the 24 hours; and the specific gravity of the urine in the bottle without the yeast is 1·042, and in the bottle with yeast 1·033, or 9 degrees less, (which represents the loss occasioned by the formation of carbonic acid) and each degree thus lost represents one grain of sugar; then 160 ounces multiplied by 9 gives 1440 grains of sugar passed in the 24 hours.

BILE. Urine containing biliary matter is generally of a deep brown yellow-colour, and imparts a yellow stain to linen rags dipped in it. It exhibits the following characteristic reactions.

1. *Test for bile pigment.* If some concentrated nitric acid containing traces of nitrous acid be placed at the bottom of a small, white, flat porcelain dish and a drop or two of urine be carefully floated over the surface of the acid, if bile pigment is present, a play of colour takes place of which green and violet is the most conspicuous. The green colour alone is only conclusive of the presence of bile, as the others may be produced by the reaction of the natural colouring matter of the urine with strong acid.

2. *Test for Bile Acids.* Dissolve a fragment of

cane sugar in as little water as possible, and carefully add some concentrated sulphuric acid drop by drop to avoid carbonizing the sugar. This is introduced into a test tube, and a few drops of urine allowed to run down the sides; at the junction of the two fluids a beautiful purple-red rim will be formed if the bile acids are present. If the colour is red or violet the reaction is due to the colouring matter of the urine, and no bile acids are present; the latter always forming a deep purple-red that cannot be mistaken.

Mr. Francis, at the laboratory of Charing Cross Hospital, has suggested the use of grape sugar instead of cane sugar, as it does not carbonize on the addition of concentrated sulphuric acid. He dissolves a few grains of grape sugar in 3i of concentrated sulphuric acid, thus forming a solution of *sulpho-saccharic acid;* this he calls "Bile Acid Test Solution."

In testing, a small quantity of urine is introduced into a test tube and a drop or two of the test solution allowed to run down the sides of the glass when at the junction of the two liquids the purple rim will be developed.

To obtain the bile acids from urine, Neukomm gives the following process. Evaporate the urine to a thick syrup, and treat with ordinary alcohol; evaporate this alcoholic solution and treat the residue with absolute alcohol. This solution is also evaporated and the residue dissolved in a little water, and the solution precipitated with neutral and basic lead acetate. The precipitate is collected, after being allowed to subside for 12 hours, and treated with sodium carbonate solution, and the solution filtered. The filtrate will contain sodium glyco-cholate and tauro-cholate, which give with Pettenkofer's test the characteristic reaction, when only ·001 per cent. is present in the urine.

URINARY SEDIMENTS.

1. *Uric Acid.* Does not dissolve when heated; insoluble in acetic acid and mineral acids; soluble in liquor potassæ; gives a purple reaction (murexide) when heated with nitric acid and ammonia.

Uric acid sediment is only found in very acid urine associated with acid sodium urate, the latter dissolves when the urine is heated. The sediment to the naked eye looks like minute grains of cayenne pepper; under the microscope the characteristic crystals are easily recognized.

2. *Urates.* Dissolve when heated; insoluble in acetic acid; soluble in alkaline solutions; give a purple reaction when heated with nitric acid and ammonia.

Sodium urate occurs as an amorphous, granular precipitate, and is only found in acid urine.

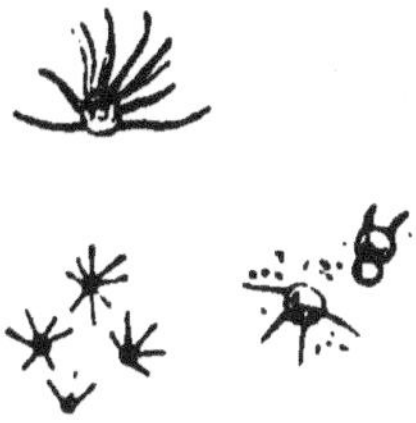

Ammonium urate is sometimes met with in alkaline urine mixed with crystals of the earthy phosphates. The crystals are known by their spiked globular form. Fig. 10. Treated with caustic potash they give off ammonia.

Fig. 10.

Calcium urate is comparatively rare; burnt on platinum foil it leaves a residue of calcium carbonate.

The colour of sediment varies, sometimes it is a yellowish white, pink, fiery or purplish red; the latter colours are generally observed in febrile states of the system.

3. *Earthy Phosphates.* Do not dissolve when heated; insoluble in alkaline solutions; soluble in acetic acid.

This sediment only occurs in alkaline urine, for so long as the urine maintains its normal acidity they are retained in solution. If therefore we add to freshly-passed urine a few drops of ammonia, a deposit of phosphates will at once be obtained. The sediment consists of calcium phosphate and ammonio-magnesium phosphate, usually together, rarely separate. The crystals of ammonio-mag-

Fig. 11. Crystals of Ammonio-
magnesium Phosphate

nesium phosphate (Fig. 11.) are either deposited as delicate feathery crystals, or as right rhombic prisms; the calcium phosphate also assume two forms, the amorphous and crystalline. The crystals

Fig. 12.
Crystallized Phosphate of Lime.

in the latter case vary considerably in form and size; appearing most frequently as thin, twisted, acicular crystals crossing each other at right angles, or thick and wedge-shaped, or small and pointed, united together to form ropes and rosettes. (Fig. 12.)

4. *Calcium Oxalate.* Insoluble in water which distinguishes it from crystals of common salt, and acetic acid. Soluble in dilute hydrochloric acid, and in strong solutions of acid sodium phosphate.

Calcium oxalate rarely appears as a sediment if the urine be very acid; if this salt is suspected to be present in acid urine it will be necessary to neutralize that fluid with ammonia. The crystals are

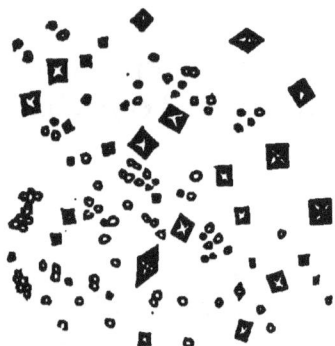

Fig. 13.   Calcium Oxalate.

recognized under the microscope as small, bril-
liànt, square octohedra, having a strong refractive
power.[*]   (Fig. 13.)

5. *Cystin* does not dissolve when heated, insoluble in acetic acid, soluble in mineral acids;
burnt in air it gives off an odour of prussic acid.

The crystals are thin, transparent,
hexagonal plates (Fig. 14.); they may
be mistaken for uric acid, but are solu
ble in strong mineral acids which the
latter are not.

Fig. 14.   Cystin.

6. *Blood.*   The urine varies in colour, from a
slight smoky tint, to a deep red, or chocolate
brown, according to the quantity of blood mingled
with it.

If the blood corpuscles are not dissolved, they
will be recognized under the microscope.

* Other varieties of crystalline form are met with as the
dumb bell, discoid, and diamond-shaped crystal, but the letter-
envelope shape of the octohedral crystals alone are characteristic.

Heat and nitric acid coagulate the albuminous matters of the blood. Heller recommends that urine containing small quantities of blood should be boiled, and concentrated caustic potash added, and again boiled. The phosphates are deposited carrying with them the colouring matter of the blood which gives them a blood-red colour. As neither bile pigment nor the colouring matter of the urine is precipitated with the phosphates, it shows that the coloration is not due to those substances.

7. *Pus.* The sediment is usually of a pale greenish-yellow. The urine contains albumin which is coagulated by heat and nitric acid. The addition of liquor potassæ converts pus into a viscid, tenacious, muco-gelatinous mass.

Under the microscope the pus corpuscles appear as minute granular vesicles, the addition of acetic acid causes them to swell up, to become more transparent, and renders the nuclei more distinctly visible.

8. *Mucus.* Does not form a true sediment, but is diffused through the urine in the form of cloudy, transparent flocculi. Mucus does not become more viscid or gelatinous on the addition of caustic alkalis; on the contrary it becomes more liquid. The addition of alcohol throws down mucin in stringy, ropy masses. Mucin differs from pyin in not being precipitable by mercuric chloride or lead acetate.

*

URINARY CALCULI.

Notice the size, colour, and general appearance; and whether on section they are made up of concentric layers, or present an uniform surface.

A portion of the stone is then broken down and reduced to powder, and if it is made up of different layers a portion of the powder of each layer must be taken, and submitted to analysis.

I. Little or no residue is left when a small portion of the powder is burnt on platinum foil, the calculus may consist of Uric Acid, Xanthin, or Cystin.

1. *Uric Acid Calculi* are the most common of all the urinary calculi, and often attain a considerable size; they are usually smooth, and of a light yellow or reddish brown colour.

The burnt powder evolves an odour of prussic acid, and of burnt animal matter; the ash is extremely small, and contains traces of sodium phosphate and carbonate.

Dissolved in liquor potassæ, and reprecipitated by the addition of hydrochloric acid, the characteristic crystals will be observed under the microscope.

Heated with nitric acid and ammonia the purple colour (murexide) will be produced.

2. *Xanthin Calculi* are extremely rare, they are usually smooth, and of a cinnamon colour, taking a polish when rubbed.

They dissolve in nitric acid without effervescence;

and do not yield the murexide reaction with nitric acid and ammonia.

3. *Cystin Calculi* are also very rare, they have a smooth surface, a greenish-yellow colour, and break with a crystalline fracture; they are the softest of all calculi, and for some days after their removal are compressible.

Cystin dissolves in the caustic alkalis and in strong mineral acids; on evaporating the ammonia solution cystin is deposited in regular hexagonal tables; on evaporating the hydrochloric solution cystin crystallizes in radiating needles.

Boiled in a solution of caustic potash with a little lead acetate, a black precipitate of lead sulphide is thrown down, which is due to the presence of sulphur contained in the cystin.

II. A residue is left when a small portion of the powder is burnt on platinum foil, the calculus may consist of Urates, Calcium and Magnesium Phosphates, Calcium Oxalate, and Calcium Carbonate.

1. *Calculi composed* entirely *of Urates* are rare, the urates being generally associated with uric acid.

The powder is treated with boilng water, which dissolves the urates leaving the uric acid; the filtrate is then evaporated and gives with nitric acid and ammonia the murexide reaction.

2. *Calculi of the Earthy Phosphates* consist of a mixture of Calcium phosphate and ammonio-magnesian phosphate; they are generally very large, of a dirty white colour, and their surface uneven.

They dissolve in hydrochloric acid without

effervescence; and under the blowpipe flame fuse, without being consumed, into a hard white mass.

Sometimes the calculus consists entirely of calcium phosphate, or of ammonio-magnesium phosphate. To discover which phosphate is present, or to separate them from each other, the calculus must be dissolved in dilute hydrochloric acid, and ammonia added drop by drop so as not completely to effect the neutralization of the mixture, then on the addition of ammonia oxalate if lime is present a precipitate of calcium oxalate is thrown down. If, on the other hand, the calculus consists of the triple phosphate no precipitate occurs till the fluid is completely neutralized with ammonia, when a crystalline precipitate consisting of stellate and prismatic crystals of ammonio-magnesium phosphate is deposited.

3. *Calcium Oxalate Calculi* are either small and pale coloured; or large, dark coloured, with a rough, irregular surface, and on section present an angular structure, with irregular, dark-coloured laminæ. The smaller calculi from their size and appearance are frequently termed " hemp-seed calculi"; the larger "mulberry calculi."

Heated on platinum foil it first of all chars from the combustion of the organic matter, and gives off an odour of burnt animal matter; finally the residue becomes white, and consists of calcium carbonate, which dissolves with effervescence on the addition of hydrochloric acid. This solution of calcium chloride when neutralized with ammonia throws down a precipitate of calcium oxalate on the addition of oxalic acid. A portion of the calculus dissolved in dilute hydrochloric acid gives a white precipitate, of ammonium oxalate, with ammonia.

4. *Calcium Carbonate Calculi* are rare, but when met with are generally multiple, occurring in large numbers in the same individual. They dissolve with effervescence in hydrochloric acid.

APPENDIX. I.

Weights, Measures and Instruments employed in the quantitative analysis of the Tissues and Fluids, and the usual method of procedure.

Chemists usually employ the Metric or French system in making their calculations. In this system the *gramme* is taken as the *unit of weight*, which represents the weight of a cubic centimetre of distilled water, at its greatest density; viz. 4° C.

The *unit of capacity* is the *litre*, which contains 1000 cubic centimetres; consequently a litre of distilled water weighed at 4°C. should weigh 1000 grammes.

The *multiples of these units* are characterized by Greek prefixes: thus,

Gramme . . .	1·		Litre	1·
Decagramme .	10·		Decalitre . . .	10·
Hectogramme .	100·		Hectolitre . .	100·
Kilogramme .	1000·		Kilolitre . . .	1000.

the unit of each denomination being ten times as great as the preceding one.

The *sub-multiples* of the units are characterized by Latin prefixes; thus,

Milligramme . .	·001		Millilitre . . .	·001
Centigramme . .	·01		Centilitre . . .	·01
Decigramme .	·10		Decilitre . . ·	·10
Gramme . . .	1·		Litre	1·

Here the unit of each denomination is ten times less than the one below it.

MEASURES OF WEIGHT.

(Dr. Warren De La Rue.)

	English grains.	Troy ounces.	Avoirdupois lbs.
Milligramme . . .	0·015432	0·000032	0·0000022
Centigramme . .	0·154323	0·000322	0·0000220
Decigramme . .	. 1·543235	0·003215	0·0002205
Gramme	15·432349	0·032151	0·0022046
Decagramme . .	154·323488	0·321507	0·0220462
Hectogramme . .	1543·234880	3·215073	0·2204621
Kilogramme . . .	15432·348800	32·150727	2·2046213

MEASURES OF CAPACITY.

	Cubic inches.	In pints.	In gallons.
Millilitre	,0·061027	0·001761	0·00022010
Centilitre	0·610271	0·017608	0·00220097
Decilitre	6·102705	0·176077	0·02200967
Litre	61·027052	1·760773	0·22009668
Decalitre	610·270515	17·607734	2·20096677
Hectolitre . . .	6102·705152	176·077341	22·00966767
Kilolitre	61027·051519	1760·773414	220·09667675

For all practical purposes, however, it is suffi-cient to remember that

1 gramme	= nearly 15·4 grains Eng :
1 kilogramme	= ,, 2 lbs. 3 oz. 5 drs.
1 litre	= ,, 1 pt. 15 fl. oz. 2 drs.
1 decilitre or 100 C.C's. =	,, 3 fl. oz. 3 drs.
1 cubic centimetre	= ,, 16·3 minims.

In making most quantitative analyses, two me-thods are usually employed; viz., *Gravimetric*, in which the substance after isolation from the mixture is weighed in the *balance*. 2. *Volumetric ;* here the substance is estimated by adding a standard solution of some chemical agent, till a definite and characteristic reaction occurs ; the number of *volumes* required to effect this reaction corresponding to the quantity of the substance present in the mixture.

The Balance usually employed is represented in fig. 15, enclosed in a case fitted with glass doors (*cc*). It carries 100 grammes, and indicates

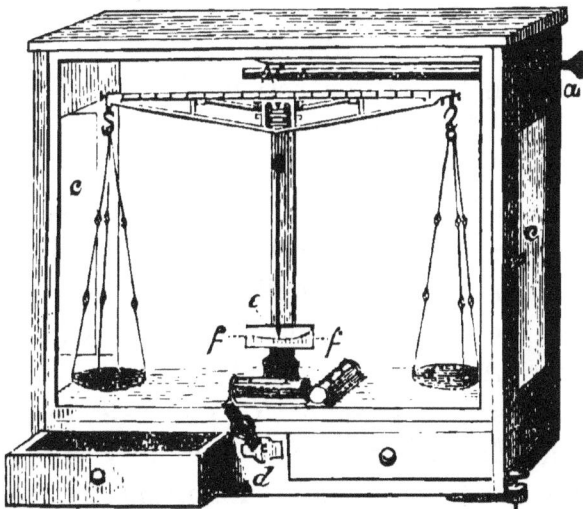

Fig. 15.

a variation of $\frac{1}{5}$ milligramme or $\frac{1}{300}$ grain, this variation is ascertained by the pointer (*e*) moving decidedly one side or other of the central mark on the register (*f*), when in equilibrium it vibrates an *equal* distance on each side of the central mark. When at rest or whilst the weights are being adjusted the instrument is kept at rest by means of a contrivance which steadies the pans ; this contrivance is worked by a screw (*d*) outside the case.

To ascertain the weight of a substance, say of a platinum capsule, we proceed as follows.

1. We ascertain if the balance is in equilibrium; this is done by liberating the pans by moving the screw (c) and observing if the pointer swings to an equal distance on each side the central mark on the register; if it does not swing to an equal distance but moves to one side more than the other, the balance must be adjusted. A brass arm about 2 inches long fixed to a screw above the fulcrum serves for this purpose; a deflection to either side makes that the heaviest.

2. The pans are steadied and the capsule, perfectly clean, and dried previously in an air bath, is placed in one pan, usually the one on the left hand of the operator, and the weights placed in the other. At first we guess at the probable weight and place say 10 grammes on the scale, this is too much, for the pointer goes to the other side of the register towards the pan supporting the capsule when we liberate the pans. We then try 5 grammes; this is too little, for on turning screw (*d*) the pointer swings towards the pan containing· the weights. We then try an intermediate weight, and add 2 grammes, which makes 7 grammes, this is not quite enough, so we add 0·5 gramme, this is too much. Take off 0·5 and try 0·2, too little; add 0·2 to make 0·4, too little; add ·05, too little; add ·02, too much; take off ·02 and try ·01, too little. Move rider by means of rod (*a*) on to

index 5 on beam, this is too much; move it to index 3, this is right! the pointer swings exactly on each side of the central mark on register. The weight therefore of the capsule is 7·463 grammes.

The Burette is employed for measuring off the volumes of standard solu-. tions required to effect the precipitation etc. of a substance from a mixture. The burette most commonly used is Mohr's fig. 16.

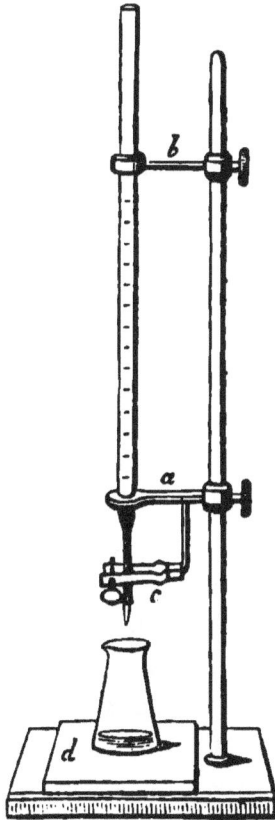

It consists of a straight tube about 18 inches in length graduated with a scale, 0° being at the top and 50° at the bottom. The tube is open at the top but at the bottom it is drawn out and contracted and fitted into a piece of caoutchouc piping at the end of which is a fine glass jet. To prevent the liquid running out, the caoutchouc piping is compressed with a pinch-cock. (*c.*) The burette is charged by pouring in the fluid at the top and filling it quite full, the pinch-cock is then pressed and a few drops allowed to escape till the *under border of the dark zone* of the fluid corresponds with 0° of the graduated scale. Fig. 17.

Fig. 16.

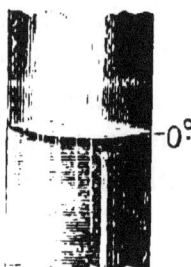

Fig. 17.

Pipettes Fig. 18, are in reality only smaller

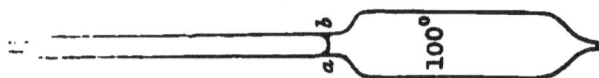

Fig. 18.

burettes, they serve to measure off a definite quantity of fluid; the most convenient sizes are 5, 10, 20 and 100 C.C., and should be made to deliver the quantity independently of what adheres to the inner surface. To use the pipette, the nozzle is dipped in the fluid and gentle suction made till the fluid rises above the mark (*a b*) in the neck. The upper opening is then closed with a moist finger, and the outside of the pipette dried, the finger is then removed and a few drops of the fluid allowed to run out, till the fluid reaches the level of the mark on the neck, the fluid thus measured can be transferred to a beaker.

The burette being charged, and the fluid for examination measured off by the pipette into a beaker, we proceed to make the analysis as follows: by pressing the pinchcock a few drops are allowed to run from the burette into the fluid in the beaker and the mixture well stirred, a drop or two of the fluid is cautiously added till some decided reaction is produced, such as change of colour, a precipitate, or reaction with test paper. When the process is complete we calculate the number of C.C.'s required to effect the reduction and make the estimation accordingly.

Thus, if 18 C.C. of the standard solution have been used, and each C.C. of this solution corresponds to ·01 of the substance present in the mixture, and 20 C.C. of the mixture have been employed, then $20 \times ·01 = ·2$, or $\dfrac{1000 \times ·2}{20} = 50$ parts in 1000.

Method of making a Quantitative Examination of a Tissue of Fluid.

Divide the substance to be examined into 2 portions.

Portion (A.) Evaporate, say 50 grms., over a water bath to ascertain the proportion of water to the total solids. (See page 100.) For ordinary purposes a cheap and efficient water bath can be made from an old oil can, by fitting it with a cork perforated to admit an ordinary glass funnel. The can is filled with water and the platinum capsule placed in the open part of the funnel, the escape of steam being allowed by a folded piece of paper placed between the edge of the filter and the capsule. The apparatus is supported on an ordinary retort stand and ebullition maintained on an argand gas lamp placed below.

A useful kind of water bath and filter is represented in Fig. 19. It can be also used as an ice

Fig. 19.

bath, by substituting ice instead of boiling water;
in this way it may be used in the preparation of
those substances that have to be isolated at 0° C.

The residue obtained by evaporation, before
weighing, must be dried in an air bath in order to
get rid of any last traces of moisture, the temperature is regulated by means of a thermometer
passing through a tube into the bath.   Fig. 20.

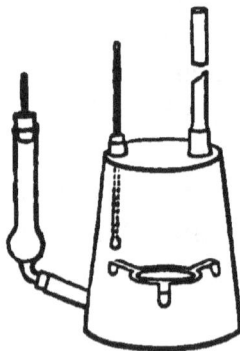

Fig. 20.

The weight of the dried residue thus obtained represents the *total solids;* viz., the organic and inorganic, the former are burnt off by using a high temperature which can be obtained by burning over a Bunsen's lamp as shown at Fig. 21. the re-

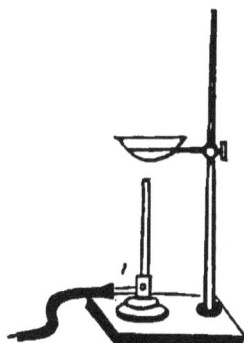

Fig. 21.

sidue left representing the *inorganic ash,*[°] which must be examined for carbonates, chlorides, phosphates, sulphates, calcium, magnesium, sodium, potassium and iron as directed under their respective headings.

Portion (B.) 50 grms. of the tissue or fluid are evaporated over the water bath to dryness, the dry residue is then transferred to a mortar, triturated, and thoroughly exhausted with boiling ether. After standing some time the etherial solution is poured off and evaporated in a weighed platinum capsule, the increase of weight of the capsule re-

* To make a thorough investigation of the inorganic ash it will be necessary to incinerate a considerable quantity of the tissue or fluid, this can be done after the proportion of inorganic matter to the organic matter and water has been first ascertained.

presenting the *fatty matters* 50 grms. of the tissue or fluid. (See also page 16.)

The residue left after exhaustion with ether is treated with boiling water for some hours, filtered, and evaporated. The residue thus obtained is treated with alcohol, this alcoholic solution contains the *alcoholic extractives*, and the residue which is insoluble in alcohol represents the *aqueous extractives*, in 50 grms. of the tissue or fluid. The extractives which generally consist of Urea, Sugar, Kreatin, Uric acid, etc., can be determined by their characteristic reactions, for this purpose it will be necessary to use a larger quantity of tissue or fluid than that employed in determining the proportion of extractives to the other substances in the mixture.

The residue which was not dissolved by the boiling water represents the *Albuminous matters* in 50 grms. plus a small quantity of inorganic matter; to obtain exactly the true amount of Albumin the residue must be incinerated and the weight of the ash deducted, (for example, see page 154.)

We have now ascertained the quantity of Water, Salts, Fatty Matters, Extractives, and Albumins in 50 grms., to estimate, these for a 1000 parts, it is only necessary to make the following calculation

$$\frac{n \times 1000}{50} = x$$ in 1000 parts; *n* representing the quantity of the substance in 50 parts.

ESTIMATION OF NITROGEN.

In some chemico-physiological enquiries, we have to determine the quantity of nitrogen ingested or egested during the 24 hours. For this purpose, the bottom of a small tubular retort is covered with soda lime, recently heated to redness, to the depth of 1·5 centimetre; and 100 C.C. of the dilute standard sulphuric acid introduced into a small flask, which is connected with the retort by a narrow bent tube. 5 C.C. of urine, for example, are poured upon the soda lime and the stopper of the retort quickly replaced, when bubbles of gas soon arise in the sulphuric acid. When the sulphuric acid begins to pass back to the neck of the retort, the retort must be gently warmed over a spirit lamp which drives back this acid. When the whole of the water is driven off from the retort the process goes on very steadily; when complete bubbles of gas are no longer disengaged, and the sulphuric acid begins to rise in the tube, the heat must be now removed, and the contents of the flask washed out into a beaker glass, a few drops of litmus being added to them, and the non-saturated sulphuric acid measured with an equivalent quantity of the soda solution.

Now 100 C.C. of sulphuric acid were employed, 30 of which are now found to be saturated, and as 1 C.C. of sulphuric acid = ·0035 gramme of nitrogen, therefore the 5 C.C. of urine contain 30 × ·0035 gramme = 0·1050 gramme of nitrogen, consequently if 1200 C.C. of urine have been passed during the 24 hours $\dfrac{0\cdot1050 \times 1200}{5} = 23\cdot2$

grammes of nitrogen in the 24 hours.

1. *Standard solution of sulphuric acid.* 12·6 grammes of oil of vitriol are weighed and diluted up to a litre of distilled water; the quantity of sulphuric acid, in each 20 C.C. should be determined by barium chloride solution, so that each 100 C.C. of the dilute acid should contain 1·0 gramme of sulphuric acid, and this corresponds to ·35 grammes of nitrogen. Consequently 1 C.C. contains ·0035 grms. of nitrogen.

2. *Standard Soda solution.* This solution must be standardized. So that an equal volume should exactly saturate the sulphuric acid.

APPENDIX II.

Standard Solutions required in Volumetric Analysis.

No. 1. *For Iron.* The solution of Potassium Permanganate. A concentrated solution of 10 parts of potassium hydrate is added to S parts of manganese peroxide, and 7 parts of potassium chlorate, the mixture evaporated to dryness, and the residue heated in a platinum crucible to redness, until the potassium chlorate is decomposed. Triturate the green mass, and boil until the green colour has changed to the violet tint of the permanganate; decant the solution, remove the precipitate, and filter through asbestos.

To determine the strength of the solution; weigh off 7·543 grms. of pure, dry, crystallized potassium ferrocyanide, corresponding with 1 grm. of iron, and dissolve in distilled water, added to fill a litre measure. 10 C.C. of this solution represents 0·010 grm. of iron.

Now take 10 C.C. of the potassium ferrocyanide solution, and dilute with 50 C.C. of distilled water acidulated with a little hydrochloric acid, place the vessel containing the solution on a sheet of white paper, and add the permanganate solution till a yellowish-red colour appears in the fluid, on agitation. If 20 C.C. of the permanganate solution is used, the graduation is complete, and 1 C.C. of it will correspond to 0·0005 grm. of iron. But if more or less than 20 C.C. are required, then the permanganate solution must be concentrated or diluted till the required strength, viz., 20 C.C., to produce the yellow-red colour with 10 C.C. of potassium ferrocyanide solution, is attained.

No. 2. *For Urea.* The solution of Mercuric Nitrate. Dissolve 77.2 grms. of pure, dry mercuric oxide in strong nitric acid, evaporate the solution to the consistence of a syrup and dilute up to 1 litre with distilled water. 1 C.C. of this solution corresponds to 0·01 grm. of urea.

No. 3. *For Phosphates.* (a) The solution of Uranum Nitrate. Dissolve 20·3 grm. of pure uranium nitrate in strong acetic acid, and dilute the solution up to 1 litre. 1 C.C. of this solution corresponds to 0·005 grm. of phosphoric acid.

(b) The solution of Sodium Acetate. To 100 grms. of sodium acetate add 100 C.C. of strong acetic acid, dilute up to 1 litre with distilled water. 5 C.C. of this solution should be added to 50 C.C. of urine, to insure the precipitation of the uranium phosphate.

No. 4. *For Chlorides.* Solution of Silver Nitrate. 20·063 grms. of pure fuzed silver nitrate are dissolved in distilled water to fill 1 litre. 1 C.C. of this solution represents 0·01 grm. of sodium chloride, and 0·006 grm. of hydrochloric acid.

No. 5. *For Sulphates.* Solution of Barium Chloride. Dissolve 30·5 grms. of dry crystallized barium chloride in distilled water to 1 litre. 1 C.C. of this solution corresponds with 0·01 grm. of sulphuric acid.

No. 6. *Estimation of Free Acid.* The solution of Sodium Hydrate. Dissolve 6·35 grms. of caustic soda in distilled water up to 1 litre. 1 C.C. of this solution corresponds to 0·01 grm. of crystallized oxalic acid.

No. 7. *For Sugar.* (a) The Copper Solution. Dissolve 34·63 grms. of crystallized cupric sulphate in distilled water, up to 1 litre. 1 C.C. of this solution corresponds to 0·005 grm. of sugar.

(b) *The Alkaline Tartrate solution.* 173 grms. of Sodium and Potassium Tartrate and 80 grms. of potassium hydrate are dissolved in water to the measure of 1 litre. 10 C.C. of this solution are to be used with every 10 C.C. of the cupric solution. No. 8. *For Kreatinin.* The solution of Zinc Chloride. Chemically pure zinc oxide is dissolved in strong hydrochloric acid, and the solution evaporated to a thick syrup in a water bath, until all the free hydrochloric acid is driven off. The residue, when cold, is dissolved in strong spirits of wine, and the solution diluted until it has the specific gravity of 1·20.°

APPENDIX III.

List of Authors referred to, and from whose works extracts have been taken.

Dr. Bence Jones. *Lectures on Pathology and Therapeutics.* 1867.

Dr. Carpenter. *Principles of Human Physiology.* Edited by Henry Power, Esq., 7th edit. 1869.

Dr. Dalton. *Treatise on Human Physiology.* New York. 5th edit. 1871.

Dr. G. Day. *Chemistry in its relation to Physiology and Medicine.* 1860.

Professor Fownes. *Manual of Elementary Chem-*

• These solutions are prepared, with great accuracy, by Messrs. Griffin and Sons, Garrick Street, Covent Garden. The chemical apparatus described in the text is also supplied by them ; the illustrations in this work being taken from their excellent manual "Chemical Handicraft."

istry: edited by HENRY WATTS, B.A., F.R.S. 11th edition. 1873.

DR. GEORGE HARLEY. *The Urine and its derangements.* 1872.

VON DR. W. KUHNE. *Treatise on Physiological Chemistry.* 1868.

PROFESSOR MILLER. *Elements of Chemistry.* Part III. 1862.

DRS. NEUBAUER and VOGEL. *Guide to the qualitative and quantitative analysis of the Urine.* Sydenham Society's Translation. 1863.

PROFESSOR ODLING. *Lectures on Animal Chemistry.* 1868.

DR. ED. PARKES. *On the Elimination of Nitrogen.* Croonian Lectures: *Lancet,* 1871. also *Proceedings of Royal Society.*

VON W. PREYER. *On Bloodcrystals.* 1871.

PROFESSOR ROSCOE. *Lessons in Elementary Chemistry.* 1869,

DR. ED. SMITH. *Cyclical changes in Health and Disease.* 1861. *Evolution of Carbonic Acid and Urea under the influence of Blood, Exertion, etc.* *Phil. Trans.*

DR. THUDICHUM. *Manual of Physiological Chemistry.* 1872.

DR. WATTS. *Dictionary of Chemistry.* Supplement. 1872. *Articles, Blood, Bile, Kreatin, and Proteids,* by PROFESSOR MICHAEL FOSTER.

Abstracts from the principal Journals of Science and Medicine; viz.,

British and Foreign Med. Chir. Review.
British Medical Journal.

The Lancet.
Medical Times and Gazette.
Journal of the Chemical Society.
Chemical News.
Philosophical Transactions. &c.

INDEX.

A.

		PAGE
Acetic acid	... ...	47
Acid albumin	... ...	35
Acidity of urine...	... ...	190
estimation of	... ...	198
Acids	... ...	44
aromatic	... ...	55
diatomic	... ...	50
dibasic	... ...	54
fatty ...	... ...	45
monatomic...	... ...	45
monobasic ...	... ...	51
Adipose tissue ...	... ...	113
Air bath ...	... ...	221
Albumin ...	... ...	28
acid ...	... ...	35
alkali	... ...	36
estimation of in urine	...	198
blood	...	154
Albuminose	... ...	38
Albuminous principles	...	24
Alcoholic extract of bile	...	138
Alkali albumin ...	... ...	36
Alkaline carbonates	... ...	101
Alkalinity of blood		
promotes oxidation	...	105
Allantoic fluid ...	...	176
Allantoin...	... ...	
Alloxan ...	... ...	61-91
Amides ...	... ...	60
Amylaceous principles	...	1
Analysis, method of procedure	214
Animal nitrogenous bases	...	60
Appendix...	... ...	214
Aqueous vapour in breath	...	179
Areolar tissue ...	... ...	112
Aromatic acids ...	... ...	55
Arsenic ...	... ...	111
Arterial blood ...	... ...	150
Axis cylinder of nerves	...	221

B.

		PAGE
Bases nitrogenous	... ...	60
Balance ...	... ...	217
Benzoic acid	... ...	55
Bile	... ...	134
acids...	... ...	139
in urine	... ...	204
pigment	... ...	141
in urine	... ...	203

		PAGE
Bilifuscin	... ...	142
Bilihumin	... ...	143
Biliprasin	... ...	143
Bilirubin ...	... ...	141
Biliverdin	... ...	143
Blood	... ...	150
arterial and venous	...	150
Blood corpuscles	... ...	155
Bone	... ...	115
Bottëger's test for sugar	...	200
Breath	... ...	178
Brücke's method of isolating	
glycogen	... ...	11
Bunsen's burner for incinerat-	
ing residues ...	... ...	223
Burette, Mohr's	... ...	219
Butyric acid	... ...	47

C.

		PAGE
Calcium carbonate	...	102
phosphate	...	104
Capric acid	... ...	48
Caproic acid	... ...	48
Carbohydrates ...	... ...	1
Carbonates, alkaline ...	...	101
Carbonic acid ...	... ...	51
determination of ...	...	103
in breath	... ...	181
loose and stable ...	...	179
Carbon monoxide hæmoglobin	160
Cartilaginous tissue ...	...	114
Casein	... ...	36-176
Cement ...	... ...	116
Cerebric acid	... ...	76
Cerebrin ...	... ...	76
Charcoal, animal preparation	
of, for filtering	... ...	97
Chlorides	... ...	106
estimation of	... ...	106
Cholesterilines ...	... ...	21
Cholesterin	... ...	19
Cholesterones ...	... ...	21
Cholesteric acid	... ...	57
Cholic acid	... ...	57
Cholin ...	... ...	66
Chondrin...	... ...	42
Chyle ...	... ...	165
Colostrum	... ...	174
Connective tissue	... ...	112
Copper ...	... ...	111

	PAGE
Cream	173
Cystin	69
sediment	208
calculi	

D.

Dalton, Dr. observations on glycogen	10
on action of gastric juice on starchy matters ...	131
influence of food on secretion of bile	135
Damaluric acid	56
Damolic acid	56
Decomposition, products of ...	44
organic acids	44
nitrogenous bases ...	60
Dental tissue	116
Dentine	116
Derived albumins	35
Detection of blood stains ...	164
Diamides, primary ...	61-77
secondary ...	61-87
Diatomic acids	50
Dibasic acids	54
Dyslysin	58

E.

Elasticin	112
Elastic tissue	42
Enamel	116
Epidermal tissue	113
Excreta, the	177
Excretin	23
Excretolic acid	23

F.

Fatty acids	45
occurrence of	46
separation of	49
Fatty degeneration	15
of muscle	120
Fatty principles	13
estimation of	16
uses of	14
non-saponifiable	19
saponifiable	13
Faeces	185
Fehling's test solution for sugar estimation	201
Fibrin	33
in blood	153
separation of	33
in chyle	165
in lymph	167

	PAGE
Fibrinoplastic substance ...	31
Fibrogen	31
Fibrous tissue	112
Fluids of mammalian ovum ...	176
Fluorine	111
Formic acid	47
Fractional distillation ...	49
Francis, Mr. bile acid test solution	204
Frankland, Prof. force values of food	122

G.

Gastric juice	128
Gelatin	41
Gelatinous principles	41
Globulin	30
Glucose	2
in urine	201
Glycerophosphoric acid	75, 125
Glycocin	62
Glycogen	9
Glycollic acid	51
Glycocholic acid	139
Grape sugar	2
estimation of, in urine ...	201
Guanin	92

H.

Hæmatin	161
Hæmatoidin	163
Hæmatoporphyrin	162
Hæmin	163
Hæmochromogen	162
Hæmoglobin	156
optical properties of ...	158
compounds of	158
Hippuric acid	69
Hydrogen-cyanide Hæmoglobin	161
Hypoxanthin	95

I.

Inorganic constituents ...	98
Inosite	6
Intestinal juice	145
digestion	146
Iron	110
estimation of	110

K.

Keratin	43
Kreatin	83

	PAGE
Kreatinin	85

L.

	PAGE
Lactic acid	51
separation of	52
fermentation·	4
Lactose	5
estimation of, in milk ...	172
Lardacein	34
Lead	111
Leucic acid	54
Leucin	63
in urine	64
Lecithin	74
Lime	109
estimation of	109
Lithofellic acid	59
Lymph	167

M.

	PAGE
Magnesium	109
estimation of	109
Mannite	1
Meconium	186
Medullary nerve substance ...	123
Metalbumin	29
Milk	171
Sugar	5
Monamides	60-62
Monatomic acids	45
Monobasic acids	50
Moore's test for sugar... ...	200
Mucin	38
Muscle clot	118
reaction	120
serum	118
sugar	6
Muscular action due to oxidation	122
Muscular exertion influence on the elimination of carbonic acid	181
Muscular tissue	117
Myelin growths	74
Myochrome	118
Myosin	31
cause of rigor mortis	32, 120

N.

	PAGE
Naurocki's experiments on frog's (kreatinin)	85
Nerve fibres	123
reaction	123
vesicles	124
Nervous tissue	123

	PAGE
Neukomm's process for obtaining bile acids from urine ...	204
Neurilemma	123
Neurin	66
Nitrogen, determination of in urine	225
Nitrogen in expired air ...	182
Nitrogenous bases	60
Nitrous dioxide hæmoglobin	161
Non-saponifiable fats	19

O.

	PAGE
Oleic acid	48
Olein	18
Oleophosphoric acid	18
Ossein	41, 115
Osseous tissue	115
Oral digestion	128
Oxalic acid	54
Oxyhæmoglobin	158

P.

	PAGE
Palmitic acid	48
Palmitin	17
Pancreatic juice	144
Pancreatin	145
Paraglobulin	31
Paralbumin	29
Parkes, Dr. observations on urea	79
Parotid saliva	127
Pepsin	129
Pettenkofer's test for ...	59, 135
bile acids	204
Phenol	56
Phosphates, alkaline	105
earthy	104
Phosphoric acid estimation of	196
Pipettes	220
Potassium, estimation of ...	108
sulphocyanate ...	127
Primary diamides ...	61, 77
Products of decomposition ...	44
Propionic acid	47
Protagon	74
Ptyalin	127
Pus	168
Pyin	39
Pyocyanin	40

Q.

	PAGE
Quantitative estimation, general plan of procedure ...	221
gravimetric	217
volumetric	217

R.

	PAGE
Reaction of bile...	134
blood	150
chyle	165
lymph	167
milk	171
muscle	120
nerve	123
saliva	127
urine	188
Reduced hæmoglobin	159
Resinous acids	57
Respiratory process, nature of	178
Rigor mortis	31, 121
Roberts' fermentation test ...	202

S.

	PAGE
Saccharine principles	1
Saliva	127
Saponifiable fats	13
Sarcin	95
Sarcolactic acid...	53
Sarcolemma	117
Sarcosin	65
Secondary diamides ...	61, 77
Seminal fluid	175
Separation of volatile fatty acids	49
solid fatty acids...	49
lactic acid ...	52
Serolin	21
Silicon	111
Sodium, estimation of ...	108
Solid tissues of the body ...	112
Spermatin	176
Standard volumetric solutions	220
Starch	8
Stearic acid	48
Stearin	17
Sweat	183
Submaxillary saliva	127
Succinic acid	55
Sugar in urine	200
Sulphates in urine	106
Syntonin	35

T.

Table of principle albuminoids	27
Tartar	128
Taurin	67
Taurocholic acid	140

	PAGE
Taurylic acid	56
Total solids, estimation of ...	100
Trommer's test	200

U.

Urates	88, 205
Urea	77
quantitative estimation of	191
Uræmia	82
Uric acid	87
quantitative estimation of	192
Urine	188
estimation of chlorides ...	197
free acid ...	198
phosphates	196
sulphates ...	197
urea ...	191
uric acid ...	192
pigment	193
Urine, morbid	198
estimation of albumin ...	199
sugar ...	201
calculi	209
sediments	205

V.

Valeric acid	47
Vinous fermentation	2-4
Vitellin	32
Volatile matter of breath ...	183

W.

Water bath	221
Water, determination of ...	100
Weights and measures ...	204
Weiss, glycogen in muscles ...	11

X.

Xanthin	93
calculus	210
Xantho proteic reaction ...	27

Y.

Yeast action on glucose ...	4
Yeast test for diabetic urine...	200

NEW BOOKS

Published and Sold by

H. K. LEWIS, 136 GOWER STREET, LONDON.

A Handbook of the Theory and Practice of
Medicine. By FREDERICK T. ROBERTS, M.D., B.Sc.,
M.R.C.P.; Assistant Physician and Teacher of Clinical
Medicine in the University College Hospital; Assistant-
Physician, Brompton Consumption Hospital, &c. Small
8vo. (*In the Press*).

The Microscope and Microscopical Technolo-
gy. A Text-book for Physicians and Students. By Dr.
HEINRICH FREY, Professor of Medicine in Zurich, Switzer-
land. Translated from the fourth and latest German
edition, and edited by G. R. CUTTER, M.D., Clinical Assistant to the New York Eye and Ear Infirmary. Illustrated
by 343 engravings on wood and containing the price lists
of the principal microscope makers of Europe and America.
Large 8vo, cloth, 25s.

A Handbook of Post-mortem Examinations
and of Morbid Anatomy. By FRANCIS DELAFIELD, M.D.,
Curator to Bellevue Hospital, Pathologist to Roosevelt
Hospital, &c. &c. 8vo, cloth, 15s.

A Treatise on Hæmophilia sometimes called
the Hereditary Hæmorrhagic Diathesis. By J. WICKHAM
LEGG, M.D., Casualty Physician to St. Bartholomew's
Hospital. Fcap. 4to, 7s 6d.

A Guide to the Examination of the Urine;
intended chiefly for Clinical Clerks and Students. By
J. WICKHAM LEGG, M.D. Third Edit., fcap. 8vo, cloth,
2s 6d.

Digitalis: its Action and its Use. An In-
quiry illustrating the Effect of Remedial Agents over Diseased Conditions of the Heart. The Hastings Prize Essay
of the British Medical Association for 1870. By J. M.
FOTHERGILL, M.D., M.R.C.P. Small 8vo, cloth, 2s 6d.

The Heart and its Diseases: with their
treatment. By J. M. FOTHERGILL, M.D., M.R.C.P. 8vo,
12s 6d.

General Surgical Pathology and Therapeu-
tics, in fifty Lectures.   A text-book for Students and
Physicians. By THEODOR BILLROTH, Professor of Surgery,
Vienna. Translated from the fourth German Edition, with
special permission of the Author, by C. E. HACKLEY,
M.A., M.D.   Large 8vo, cloth, 18s.

A Hand-Book of Therapeutics. By Sydney
RINGER, M.D., Professor of Therapeutics in University
College.   Third edition, revised and enlarged.   Small
8vo, 12s 6d.
" This treatise on Therapeutics at once assumed the first position among
kindred works on the subject in the English language. . . . . In conclu-
sion, we have great pleasure in recommending this treatise."—*Brit. and
For. Med.-Chir. Review,* January, 1872.

On the Principal Varieties of Pulmonary
Consumption, with Practical Comments on Diagnosis,
Prognosis and Treatment.   By R. D. POWELL, M.D.
Lond.; Member of the Royal College of Physicians, London; Senior Assistant Physician to the Hospital for Consumption and Diseases of the Chest, Brompton; Lecturer
on Materia Medica at the Charing Cross School of Medicine and Assistant Physician to the Hospital.   Sm. 8vo, ∙
cloth, 3s. 6d.

A Text Book of Practical Medicine, with
particular reference to Physiology and Pathological Ana- ∙
tomy. By Dr. FELIX von NIEMEYER. Translated from the
Eighth German Edition, by special permission of the
Author, by GEORGE H. HUMPHREYS, M.D., and CHARLES
E. HACKLEY, M.D.   2 vols. large 8vo, 36s.

On Diet and Regimen in Sickness and Health,
and on the Interdependence and Prevention of Diseases
∙ and the Diminution of their Fatality. By HORACE DOBELL,
M.D., Senior Physician to the Royal Hospital for Diseases ∙
of the Chest, &c.   Fifth, revised and enlarged edition, sm.
8vo, cloth, 5s.

Affections of the Heart, and in its neighbour-
hood. Cases, Aphorisms, and Commentaries. By HORACE
DOBELL, M.D., &c.   Illustrated by the Heliotype process.
8vo, cloth, 6s. 6d.

The Diseases of the Ear : Their Diagnosis
and Treatment. By J. TOYNBEE, F.R.S. With Supplement by JAMES HINTON, M.R.C.S., Aural Surgeon to Guy's Hospital. Demy 8vo, 8s 6d.

Electricity in its relation to Practical Medicine.
By Dr. MORITZ MEYER. Translated from the third German Edition, with notes and additions by WILLIAM A. HAMMOND, M.D. With illustrations, 500 pages, large 8vo, cloth, 18s.

The Practical Medicine of To-day : Two
Addresses delivered before the British Medical Association, and the Epidemiological Society. By Sir W. JENNER, Bart., M.D. Small 8vo, cloth, 1s 6d.

Ovarian Tumours : their Pathology, Diagnosis
and Treatment, especially by Ovariotomy. By E. RANDOLPH PEASLEE, M.D., LL.D. ; Professor of Gynæcology in the Medical Department of Dartmouth College; Attending Surgeon of the New York State Woman's Hospital; President of the New York Academy of Medicine, &c. &c. Illustrations, roy. 8vo, cloth, 16s.

The Treatment of Syphilis with Subcutaneous
Sublimate Injections. By Dr. GEORGE LEWIN, Professor at the Fr. Wilh. University, and Surgeon-in-Chief of the Syphilitic Wards and Skin Diseases of the Charité Hospital, Berlin. Translated by Dr. CARL PRŒGLE, and Dr. E. H. GALE, late Surgeon United States Army. Small 8vo, 10s.

Hysterology; a Treatise, Descriptive and
Clinical, on the Diseases and Displacements of the Uterus. By EDWIN NESBIT CHAPMAN, M.A., M.D., late Professor of Obstetrics, Diseases of Women and Children, and Clinical Midwifery in the Long Island College Hospital. Illustrations, roy. 8vo cloth, 18s.

A Practical Treatise on the Diseases of
Children. By Dr. A. VOGEL. Translated and Edited by H. RAPHAEL, M.D. From the fourth German Edition, illustrated by six lithographic plates, part coloured, large 8vo, 600 pages, 18s.

On the Preservation of Health; or, essays
explanatory of the principles to be adopted by those who
desire to avoid disease. By Dr. T. INMAN. Third edition,
demy 8vo, 6s 6d.

On the Restoration of Health; being essays
on the principles upon which the treatment of many diseases is to be conducted. By Dr. T. INMAN, late Physician to the Royal Infirmary, Liverpool, &c. New edition,
enlarged, small 8vo, cloth, 7s 6d.

Human Anatomy: forming a complete series
of Questions and Answers for the use of Medical Students.
By M. REDMAN.   2 vols., 12mo, cloth, 10s 6d.

Synopsis of the British Flora.   Arranged
according to the natural orders. Containing vasculares or
flowering plants. By JOHN LINDLEY, Ph.D., F.R.S., late
Professor of Botany in University College. Third edition,
12mo, 4s.

A Practical Treatise on the Diseases of
Children. By Dr. J. FORSYTH MEIGS, M.D., Fellow of
the College of Physicians of Philadelphia, &c. &c., and
WILLIAM PEPPER, M.D., Physician to the Philadelphia
Hospital, &c., &c. Fourth Edition, roy. 8vo, cloth, cut
edges, pp. 920, 26s.

Cazeaux's Obstetrics: A Theoretical and
Practical Treatise on Midwifery including the Diseases
of Pregnancy and Parturition. Revised and Annotat-
ed by S. TARNIER. Translated from the Seventh French
Edition by BULLOCK. Royal 8vo, over 1100 pages. 172
Illustrations. 30s.

" It is unquestionably a work of the highest excellence, rich in informa-
ion, and perhaps fuller in details than any text-book with which we are
acquainted. The author has not merely treated of every question which
relates to the business of parturition but he has done so with judgment
and ability."—*Brit. Med.-Chir. Review.*

" M. Cazeaux's book is the most complete we have ever seen upon the
subject."—*N. A. ed.-Chir. Review.*

The Membrana Tympani in Health and Dis-
ease, Illustrated by 24 Chromolithographs. Clinical Con-
tributions to the Diagnosis and Treatment of Diseases of
the Ear, with Supplement. By Dr. A. POLITZER. Translated by Dr. MATTHEW, sen., and Dr. NEWTON. Large
8vo, 15s.